FIT SIGMA

For other titles in the Wiley Finance series
please see www.wiley.com/finance

FIT SIGMA

A Lean Approach to Building Sustainable Quality Beyond Six Sigma

Ron Basu

WILEY
A John Wiley and Sons, Ltd., Publication

This edition first published 2011

Registered office
John Wiley & Sons Ltd, The Atrium, Southern Gate, Chichester, West Sussex, PO19 8SQ, United Kingdom

For details of our global editorial offices, for customer services and for information about how to apply for permission to reuse the copyright material in this book please see our website at www.wiley.com.

Library of Congress Cataloging-in-Publication Data

Basu, Ron.
Fit Sigma: a lean approach to building sustainable quality beyond Six Sigma / Ron Basu.
p. cm.
ISBN 978-0-470-66621-0
1. Manufacturing processes – Quality Control. 2. Industrial Management – Quality Control. 3. Six Sigma (Quality Control Standard) I. Title.
TS156.B4326 2011
658.4′013–dc22

A catalogue record for this book is available from the British Library.

ISBN 978-0-470-66621-0 (hard back); ISBN 978-1-119-99112-0 (ebk);
ISBN 978-1-119-97372-0 (epub); ISBN 978-1-119-97373-7 (emobi)

Typeset in 10/12pt Times by Aptara Inc., New Delhi, India
Printed and bound in Great Britain by CPI Antony Rowe, Chippenham, Wiltshire

To Moira, Bonnie and Robi

Contents

Preface

BACKGROUND

Whilst passing through Miami Airport en route to Mexico City, I was killing time by perusing the newspaper, and I came across an article on 'Six Sigma' in *USA Today* which caught my eye. Dated 21 July 1998, it had a challenging message relating to the cost benefits of Six Sigma. In the current global economic downturn, this challenge is even more prevalent. At that time I was co-ordinating a global MRP(II) programme between all the manufacturing sites of GlaxoWellcome, including the Xochimilco plant in Mexico. The Global Manufacturing and Supply Division of GlaxoWellcome was considering a 'Lean Sigma' initiative which was meant to be a hybrid of Six Sigma and Lean Manufacturing. It struck me that the message in *USA Today* reflected not just the doubts (or expectations) in the minds of my colleagues but perhaps those of quality practitioners world-wide.

These doubts or expectations addressed many questions. Isn't Six Sigma simply another fad or even just a repackaged form of TQM? It appears to be successful in large organisations like Motorola and General Electric, but can small firms support such a programme? How can we apply the Six Sigma methodology originated from manufacturing operations to the far larger market of the service sector? Like any good product, Six Sigma will have a finite life cycle – so what is next? Surely one big question must be: how can we sustain the benefits in the longer term? It is good to be 'lean', but isn't it better to be 'fit' in order to stay agile?

The idea of writing a book to address these issues was mentally conceived at Miami Airport and the concept of FIT SIGMA™ was born. The resulting volume was named *Quality Beyond Six Sigma* and was published by Butterworth-Heinemann in January 2003. In the experience of the author, both in industry and management courses, the details of various tools and techniques had to be acquired from different books, publications and training manuals. It is time now to produce a comprehensive, user-friendly and 'hands-on' study. This could act as a single-source reference point for tools and techniques to be used by all practitioners and students of operational excellence. With these thoughts in mind I wrote *Implementing Quality* (published by Thomson), which was followed by *Implementing Six Sigma and Lean* (published by Elsevier). The current volume, *FIT SIGMA*, has been produced with the aim of updating *Quality Beyond Six Sigma*.

WHY FIT SIGMA?

Six Sigma and Lean Sigma may be viewed by some as 'old hat'. However, in the current global economic downturn, many organisations, including the public sector, are looking beyond traditional cost-cutting to find longer-term solutions to their financial and organisational challenges.

This near perfection for many may seem to be overkill, while to some it constitutes an impossible ideal to strive towards – especially for a small business. Many businesses are hesitant to dive into an apparently expensive Lean Sigma programme which is known to be successful for large manufacturing organisations like GE, Motorola and Dow Chemicals. There are also new challenges in many areas, including climate change, Green supply chain, emerging markets and the growing service sector. FIT SIGMA, which is an extension of Lean Sigma, is aimed to address these challenges. The methodology of FIT SIGMA – or FIT Σ – has three elements: fitness for the purpose, fitness for improvement and integration, and fitness for sustainability.

ABOUT THIS BOOK

The major new features of this book as compared to *Quality Beyond Six Sigma* include either new or rewritten chapters as follows:

Chapter 1: The Evolution of Six Sigma, Lean Sigma and FIT SIGMA™
Chapter 2: More about FIT SIGMA
Chapter 3: DMAIC Methodology for FIT SIGMA
Chapter 4: FIT SIGMA Tools
Chapter 5: FIT SIGMA in Large Manufacturing Operations
Chapter 6: FIT SIGMA in Service Operations
Chapter 7: Six Sigma in Small and Medium Enterprises
Chapter 8: FIT SIGMA in Project Management
Chapter 9: FIT SIGMA in Green Thinking

This book is a practical guide to FIT SIGMA which is necessary in the implementation of all kinds of quality programmes at whatever level or under whichever banner these may be pursued, be it continuous improvement, TQM, Six Sigma, Lean Enterprise, Lean Sigma or Lean Six Sigma, to name but a few.

WHO SHOULD USE THIS BOOK?

This volume is aimed at a broad cross-section of readership.

- Functional managers, participants and practitioners in TQM, Six Sigma and operational excellence will find that this book will provide them with a comprehensive insight into the tools and techniques of continuous improvement in a single package. A step-by-step guide is included for the application of the appropriate tools to their improvement processes. This volume could be used as an essential handbook for all employees in a Six Sigma programme.
- Senior executives, both in the manufacturing and service industries (regardless of function), will find that this book will give them a better understanding of basic tools and techniques

and help them to support a quality improvement initiative and sustain a strong competitive position.

- Professional management and training consultants will find the comprehensive approach of tools and techniques forms an essential handbook for Six Sigma-related assignments and seminars.
- Management schools and academies and research associations will find this book valuable to fill the visible gap in the basics of operational excellence. This text will provide support to both undergraduate and postgraduate courses containing quality and operational excellence and can serve as a main text book for the quality elective component for MBA students.

The readership will be global and particularly cover North America, the UK, Continental Europe, Australia and the Asia-Pacific countries.

HOW TO USE THIS BOOK

The book allows maximum flexibility for readers and for practical usage depending on individual requirements and interests. The application areas of the study include the following.

Implementing FIT SIGMA

All organisations, whether service or manufacturing, private or public sector, large, medium or small, should particularly benefit from the specific chapters for large manufacturing organisations, service organisations, small and medium organisations, projects and Green thinking, as well as the section concerning steps of implementation in Chapter 10. The programme members and task groups should acquire a copy of the book and also gain a common understanding of tools as described in Chapter 4.

University and College Courses

It can be used as a textbook or reference tool for advanced programmes on managing quality in universities and business schools. The questions and exercises included in Appendix I should help students by allowing them the chance to practice and assess their level of understanding gained from the relevant chapters. Tutors will have the opportunity of applying these questions and exercises as part of their lecture material and course content.

Enhancing Knowledge

The book contains both the strategic approach of implementing FIT SIGMA and a detailed coverage of tools and DMAIC processes which underpin the programme. The reader, whether a CEO, employee or student, should find the book acts as a self-help method of enhancing his or her knowledge and understanding of quality and operational excellence.

ADDITIONAL FEATURES

Summary

At the end of each chapter the reader will find a summary of the key elements covered in that chapter. This should provide a sound basis for general revision.

Case Studies

Case studies have been included to encourage readers to respond in the context of practical situations. These are concisely written and provide a good learning resource for tutorials.

Statistics

Efforts have been made to minimise the fear of advanced statistics in FIT SIGMA! However, Appendix II illustrates the application of basic statistics in Six Sigma and FIT SIGMA.

Glossary

A comprehensive Glossary of relevant terms has been provided at the end of the book. This gives a ready reckoner for the common terminology and phrases experienced in managing quality.

I have made every effort to furnish you with both simple and more complex concepts which are nonetheless easy to understand. I feel confident that with enough common sense you can apply them readily to make FIT SIGMA a reality in your organisations and programmes. This is intended to offer you 'more for less' through the application of FIT SIGMA; that is, more results with less effort in your change programmes. Mahatma Gandhi once said, 'Be the change you want to be in the world'. I hope this book will in some way help you to bring about that change.

Acknowledgements

I acknowledge the help and support from my colleagues and students at Henley Business School in England and SKEMA Business School in France. As always, it has been my pleasure to work with Dr Nevan Wright, my co-author of *Quality Beyond Six Sigma* and other books.

I am grateful to the many contributors to the case examples included in the book, with special mentions to Alex Morton, Andy Liddle, Mike Waterhouse and John Barnes.

Every effort has been made to credit the authors, publishers and websites of material used in this book. I apologise if inadvertently any sources remain unacknowledged and if known I shall be pleased to credit them in the next edition.

My sincere thanks go to the staff of my publishers, especially to Jenny McCall and Gemma Valler for getting this project off the ground.

Finally, the project could not have been completed without the encouragement and help of my family, especially my wife Moira and daughter Bonnie.

About the Author

Ron Basu is Director of Performance Excellence Limited and a Visiting Fellow at Henley Business School, England. He is also a Visiting Professor at SKEMA Business School, France. He specialises in Operational Excellence and Supply Chain Management and has research interests in Performance Management and Project Management.

Previously he held senior management roles in blue-chip companies like GSK, Glaxo Wellcome and Unilever, and led global initiatives and projects in Six Sigma, ERP/MRP(II), Supply Chain Re-engineering and Total Productive Maintenance. Prior to this he worked as a Management Consultant with A.T. Kearney.

He is the co-author of *Total Manufacturing Solutions*, *Quality Beyond Six Sigma*, *Total Operations Solutions* and *Total Supply Chain Management*, and the author of books with titles *Measuring e-Business Performance*, *Implementing Quality* and *Implementing Six Sigma and Lean*. He has authored a number of papers in the operational excellence and performance management fields. He is a regular presenter of papers in global seminars on e-Business, Six Sigma and Manufacturing and Supply Chain topics.

After graduating in Manufacturing Engineering from UMIST, Manchester, Ron obtained an MSc in Operational Research from Strathclyde University, Glasgow. He has also been awarded a PhD at Reading University. He is a Fellow of the Institution of Mechanical Engineers, the Institute of Business Consultancy, the Association for Project Management and the Chartered Quality Institute. He is also the winner of an APM Project Management Award.

Ron lives with his wife Moira in Gerrards Cross, England and has two children, Bonnie and Robi.

1 The Evolution of Six Sigma, Lean Sigma and FIT SIGMA™

1.1 INTRODUCTION

> Today, depending on whom you listen to, Six Sigma is either a revolution slashing trillions of dollars from Corporate inefficiency, or it's the most maddening management fad yet devised to keep front-line workers too busy collecting data to do their jobs.
>
> *USA Today*, 21 July 1998

At the time of writing, it has been 12 years since the above statement was made. During this time the 'Six Sigma revolution' has created a huge impact in the field of operational excellence, yet conflicting views are still prevalent.

Let us evaluate the arguments for both sides. On a positive note, the success of 'Six Sigma' in General Electric (GE) under the leadership of Jack Welch is undisputed. In the GE company report of 2000, their CEO was unstinting in his praise: 'Six Sigma has galvanised our company with an intensity the likes of which I have never seen in my 40 years at GE.' Even financial analysts and investment bankers compliment the success of Six Sigma at GE. An analyst at Morgan Stanley Dean Witter recently estimated that GE's gross annual benefit from Six Sigma could reach 5% of sales and that share values might increase by between 10% and 15%.

However, the situation is more complex than such predictions would suggest. In spite of the demonstrated benefits of many improvement techniques such as total quality management (TQM), business process re-engineering and Six Sigma, most attempts by companies to use them have ended in failure (Easton and Jarrell, 1998). Sterman *et al.* (1997) conclude that companies have found it extremely difficult to sustain even initially successful process improvement initiatives. Yet more puzzling is the fact that successful improvement programmes have sometimes led to declining business performance, causing lay-offs and low employee morale. Motorola, the originator of Six Sigma, announced in 1998 that its second-quarter profit was almost non-existent and that consequently it was cutting 15,000 of its 150,000 jobs.

To counter heavyweight enthusiasts like Jack Welch (GE) and Larry Bossidy (Allied Signal), there are sharp critics of Six Sigma. In fact, Six Sigma may sound new, but critics say that it is really just statistical process control in new clothing. Others dismiss it as another transitory management fad that will soon pass.

It is evident that like any good product, Six Sigma should also have a finite life cycle. In addition, business managers can be forgiven if they are often confused by the grey areas of distinction between quality initiatives such as TQM, Six Sigma and Lean Sigma.

Against this background, let us examine the evolution of total quality improvement processes (or in a broader sense, operational excellence) from ad-hoc upgrading, working up to TQM and then to Six Sigma and finally to Lean Sigma. Building on the success factors of these processes, the vital question is: how do we sustain the results? The author has named this sustainable process FIT SIGMA^{TM} (see Basu and Wright, 2003).

So, what is FIT SIGMA? Firstly, take the key ingredient of quality, then add accuracy in the order of 3.4 defects in 1,000,000. Now implement this across your business with an intensive education and training programme. The result is Six Sigma. Now let's look at Lean Enterprise, an updated version of classical industrial engineering. It focuses on delivered value from a customer's perspective and strives to eliminate all non-value-added activities ('waste') for each product or service along a value chain. The integration of the complementary approaches of Six Sigma and Lean Enterprise is known as Lean Sigma. FIT SIGMA is simply the next wave. If Lean Sigma provides agility and efficiency, then FIT SIGMA allows a sustainable fitness. In addition, the control of variation from the mean (small sigma 'σ') in the Six Sigma process is transformed to company-wide integration (capital Sigma 'Σ') in the FIT SIGMA process. Furthermore, the philosophy of FIT SIGMA should ensure that it is indeed fit for the organisation.

The road map to FIT SIGMA (see Figure 1.1) contains three waves and the entry point of each organisation will vary:

- First Wave: As is to TQM.
- Second Wave: TQM to Lean Sigma.
- Third Wave: Lean Sigma to FIT SIGMA.

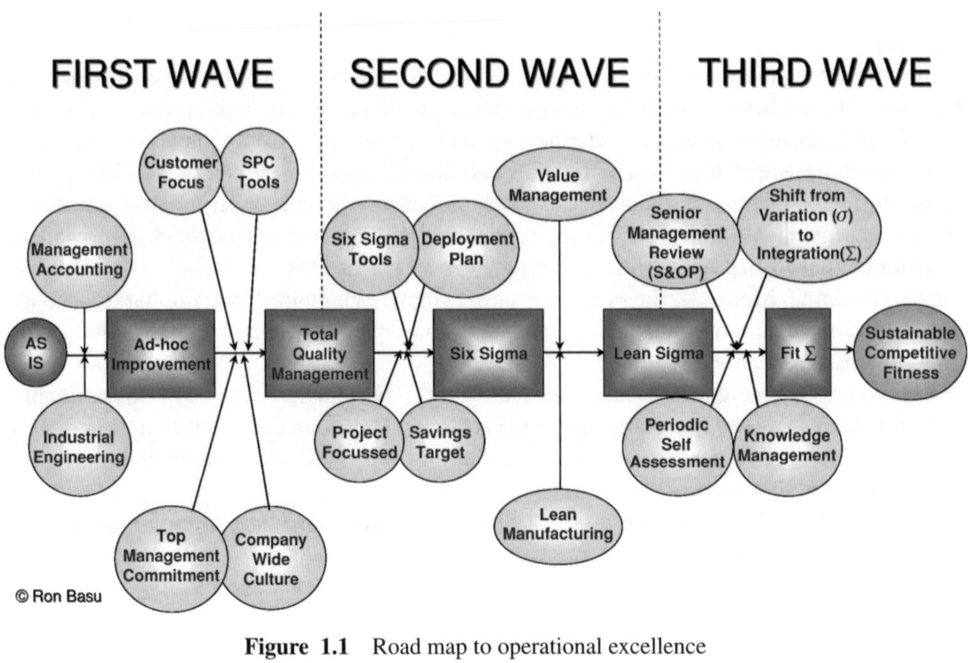

Figure 1.1 Road map to operational excellence

1.2 FIRST WAVE: AS IS TO TQM

The organised division of labour to improve operations may have started with Adam Smith in 1776. However, it is often the industrial engineering approach, which has roots in F.W. Taylor's 'Scientific Management', that is credited with the formal initiation of the first wave of operational excellence. This industrial engineering approach was sharpened by operational research and complemented by operational tools such as management accounting.

During the years following the Second World War, the 'First Wave' saw through the rapid growth of industrialisation; but in the short term the focus seemed to be upon both increasing volume and reducing the cost. In general, improvement processes were 'ad-hoc', factory-centric and conducive to 'pockets of excellence'. Then in the 1970s the holistic approach of TQM initiated the 'Second Wave' of operational excellence. The traditional factors of quality control and quality assurance are aimed at achieving an agreed and consistent level of quality. However, TQM goes far beyond mere conformity to standard. TQM is a company-wide programme and requires a culture in which every member of the organisation believes that not a single day should go by within that organisation without in some way improving the quality of its goods and services.

1.3 SECOND WAVE: TQM TO LEAN SIGMA

Learning the basics from W.E. Deming and J.M. Juran, Japanese companies extended and customised the integrated approach and culture of TQM (Basu and Wright, 1998; Oakland, 2003). Arguably, the economic growth and manufacturing dominance of Japanese industries in the 1980s can be attributed to the successful application of TQM in Japan. The three fundamental tenets of Juran's TQM process are firstly, upper management leadership of quality; secondly, continuous education on quality for all; and finally, an annual plan for quality improvement and cost reduction. These foundations are still valid today and embedded within the Six Sigma/Lean Sigma philosophies. Phil Crosby and other leading TQM consultants incorporated customer focus and Deming's SPC tools and propagated the TQM philosophy both to the USA and the industrialised world. The Malcolm Baldridge Quality Award, ISO 9000 and the Deming Quality Award have enhanced the popularity of TQM throughout the world, while in Europe the EFQM (European Foundation of Quality Management) was formed. During the 1980s, TQM seemed to be everywhere and some of its definitions – such as 'fitness for the purpose', 'quality is what the customer wants' and 'getting it right first time' – became so over-used that they were almost clichés. Thus the impact of TQM began to diminish.

In order to complement the gaps of TQM in specific areas of operation excellence, high-profile consultants marketed mostly Japanese practices in the form of a host of three-letter acronyms (TLAs), such as JIT, TPM, BPR and MRP(II). Total productive maintenance (TPM) has demonstrated successes outside Japan by focusing on increasing the capacity of individual processes. TQM was the buzzword of the 1980s but it is viewed by many, especially in the US quality field, as an embarrassing failure – a quality concept that promised more than it could deliver. Phil Crosby pinpoints the cause of TQM 'failures' as 'TQM never did anything to define quality, which is conformance to standards'. Perhaps the pendulum swung too far towards the concept of quality as 'goodness' and the employee culture. It was against this background that the scene for Six Sigma appeared to establish itself.

Six Sigma began back in 1985 when Bill Smith, an engineer at Motorola, came up with the idea of inserting hard-nosed statistics into the blurred philosophy of quality. In statistical terms, sigma (σ) is a measure of variation from the mean; thus the greater the value of sigma, the fewer the defects. Most companies produce results which are at best around four sigma, or more than 6000 defects. By contrast, at the six sigma level, the expectation is only 3.4 defects per million as companies move towards this higher level of performance.

Although invented at Motorola, Six Sigma has been experimented with by Allied Signal and perfected at General Electric. Following the recent merger of these two companies, GE is truly the home of Six Sigma. During the last five years, Six Sigma has taken the quantum leap into operational excellence in many blue-chip companies including DuPont, Ratheon, Ivensys, Marconi, Bombardier Shorts, Seagate Technology and GlaxoSmithKline.

The key success factors differentiating Six Sigma from TQM are:

1. The emphasis on statistical science and measurement.
2. A rigorous and structured training deployment plan (Champion, Master Black Belt, Black Belt and Green Belt).
3. A project-focused approach with a single set of problem-solving techniques such as DMAIC (Define, Measure, Analyse, Improve, Control).
4. Reinforcement of the Juran tenets (Top Management Leadership, Continuous Education and Annual Savings Plan).

Following their recent application in companies like GlaxoSmithKline, Ratheon, Ivensys and Seagate, the Six Sigma programmes have moved towards the Lean Sigma philosophy, which integrates Six Sigma with the complementary approach of Lean Enterprise. Lean focuses the company's resources and its suppliers on the delivered value from the customer's perspective. Lean Enterprise begins with Lean Production, the concept of waste reduction developed from industrial engineering principles and refined by Toyota. It expands upon these principles to engage all support partners and customers along the value stream. Common goals to both Six Sigma and Lean Sigma are the elimination of waste and the improvement of process capability. The industrial engineering tools of Lean Enterprise complement the science of the statistical processes of Six Sigma. It is the integration of these tools in Lean Sigma that provides an operational excellence methodology capable of addressing the entire value delivery system.

1.4 THIRD WAVE: LEAN SIGMA TO FIT SIGMA

Lean Sigma is the beginning of the 'Third Wave'. The predictable Six Sigma precisions combined with the speed and agility of Lean produces definitive solutions for better, faster and cheaper business processes. Through the systematic identification and eradication of non-value-added activities, optimum value flow is achieved, cycle times are reduced and defects eliminated.

The dramatic bottom line results, and extensive training deployment of Six Sigma and Lean Sigma must be sustained with additional features for securing the longer-term competitive advantage of a company. The process to do just that is FIT SIGMA. The best practices of Six Sigma, Lean Sigma and other proven operational excellence best practices underpin the basic building blocks of FIT SIGMA.

Four additional features are embedded in the Lean Sigma philosophy to create FIT SIGMA. These are:

1. A formal Senior Management Review process at regular intervals, similar to the Sales and Operational Planning process.
2. Periodic self-assessment with a structured checklist which is formalised by a certification or award, similar to the EFQM award but with more emphasis on self-assessment.
3. A continuous learning and knowledge management programme.
4. The extension of the programme across the whole business, with the shifting of the theme of the variation control (σ) of Six Sigma to the integration of a seamless organisation (Σ).

1.5 MORE ABOUT SIX SIGMA

Six Sigma is an approach that takes a whole-system attitude to the improvement of quality and customer service so as to enhance the bottom line. The Six Sigma concept matured between 1985 and 1986, and grew out of various quality initiatives at Motorola. Like most such quality initiatives since the days of Dr Deming in the 1960s, and in particular the concept of TQM, Six Sigma requires a total culture throughout an organisation. This means that everyone, at all levels, should possess a passion for continuous improvement with the ultimate aim of achieving virtual perfection. The difference with Six Sigma is the setting of a performance level that equates to 3.4 defects per 1 million opportunities. To ascertain whether Six Sigma has been achieved requires a common language throughout the organisation (at all levels and within each function) and standardised, uniform measurement techniques of quality. The overall Six Sigma philosophy has a goal of **total** customer satisfaction.

A survey (Basu and Wright, 2003) was conducted with the following leading companies in the UK who had adopted the Six Sigma approach to quality:

- Motorola.
- Allied Signal (Honeywell).
- General Electric.
- Raytheon.
- DuPont Teijn.
- Bombadier Shorts.
- Seagate Technology.
- Foxboro (Invensys).
- Norando.
- Ericson.

The results indicated that the main driver leading to the application of Six Sigma within a company is cost savings rather than customer satisfaction! In coming to this conclusion, the firms benefited from informal networking with members of the above companies as well as leading consulting groups such as Air Academy Associates, Rath and Strong, Price Waterhouse Cooper, Iomega and Cambridge Management Consulting. The surveyed companies reported between them a long list of intangible and indirect benefits. However, these plus points did not seem to be supported by any employee or customer surveys.

Nonetheless, very real results from the adoption of Six Sigma continue to be reported. For example, in 1997 Citibank undertook a Six Sigma initiative and after just three years it was reported that defects had reduced by 10 times (see Erwin and Douglas, 2000 for details). Likewise, General Electric state that the initial $300 million invested in 1997 in Six Sigma will deliver between $400 million and $500 million savings with additional incremental margins of $100 to $200 million. Wipro Corporation in India says that from a start in 1999, after just two years defects were reduced to such an extent as to realise a gain of eight times over their initial investment in Six Sigma.

The application of operational excellence concepts is now extended to non-manufacturing processes. 'Firms such as Motorola [and] General Electric ... successfully implemented Six Sigma. Motorola saved $15 billion in an 11 year period. General Electric saved $2 billion in 1999 alone ... Although Six Sigma initiatives have focused primarily on improving the performance of manufacturing processes, the concepts are widely applied in non-manufacturing, administrative and service functions' (Weinstein *et al.*, 2008).

1.6 WHAT IS SIX SIGMA?

So just what is the enigma of Six Sigma? Sigma is a classical Greek letter (σ), which is used in mathematical and statistical models to signify the standard deviation from the mean. This might sound like statistical mumbo-jumbo, but in reality it is a very simple concept. The mean (more correctly referred to as the arithmetic mean) is what most of us would call the average. For example, if a cricket player batted 10 times and the total of his or her 10 scores is 650, then the average is 65 (even though s/he might have 'scored' nil on one occasion and 250 on another). This is because mathematically, each turn to bat contributes to the average. In statistical terms, the arithmetic mean of the total score of 650 is 65, arrived at by dividing the total number of runs by the number of innings (650/10).

The next basic concept in statistics is frequency distribution. An often-quoted example in statistical textbooks is the tossing of 10 coins 100 times. The result of each throw of the 10 coins could range from 10 heads and no tails, to 10 tails and no heads, or any combination in between – i.e. one head and nine tails, two heads and eight tails, and so on. We would expect that if the coins are evenly balanced then we are more likely to have a probability of five heads and five tails than we are to get 10 heads and no tails!

Table 1.1 shows the result of tossing 10 coins 100 times. This can be shown as a histogram (see Figure 1.2) and also as a distribution curve (see Figure 1.3).

The curve shown in Figure 1.3 is an example of a normal distribution curve. The curve is bell-shaped (i.e. it is symmetrical from the midpoint). Of course not all distributions will give this outline, but under normal circumstances given a large enough population, in our example concerning 100 throws of 10 coins, it is very likely that the distribution curve will be similar to that shown in Figure 1.3. The midpoint is shown on our curve as 'x'. In statistical language, x represents the measure of central dispersion but in everyday English the term 'midpoint' indicates the same thing and is good enough for us!

If we assume a normal distribution curve as shown in Figure 1.3, one standard deviation from both sides of the midpoint (midpoint plus or minus one sigma) will include 68.27% of the total, and two standard deviations (two sigma) from both sides of the midpoint will include 95.45% of the total. Thus three standard deviations (three sigma) will cover 99.73%. If we extend out to six standard deviations (six sigma) from each side of the midpoint we cover 99.99966% of the total!

Table 1.1 Result of tossing 10 coins 100 times

Number of heads	0	1	2	3	4	5	6	7	8	9	10
Frequency	1	2	5	12	18	23	16	10	9	3	1

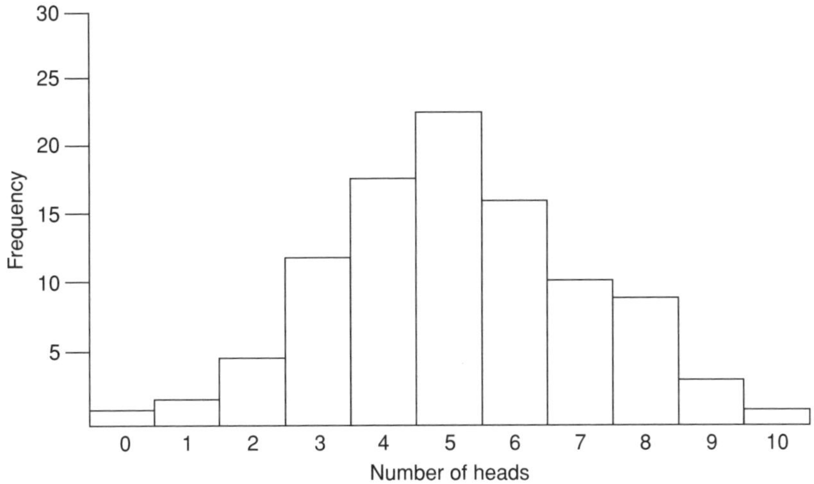

Figure 1.2 A histogram of tossing 10 coins

Now let us look at the quality programme known as Six Sigma, our subject. Regarding this we can observe that for a process, the higher the sigma the more the outputs of the process (be they products or services) and the closer they will be to always meeting the customer requirements. In other words, the higher the sigma the fewer the defects. It should be noted that the higher-multiple sigma does not increase the 'variation', it only increases the area covered under the distribution curve. For example:

- With one sigma, 68.27% of products or services will meet customer requirements and there will be 317,300 defects per million opportunities.
- Whereas with three sigma, 99.73% of products or services will meet customer requirements and there will be 2700 defects per million opportunities.
- And with six sigma, 99.99966% of products or services will meet customer requirements and there will be 3.4 defects per million opportunities.

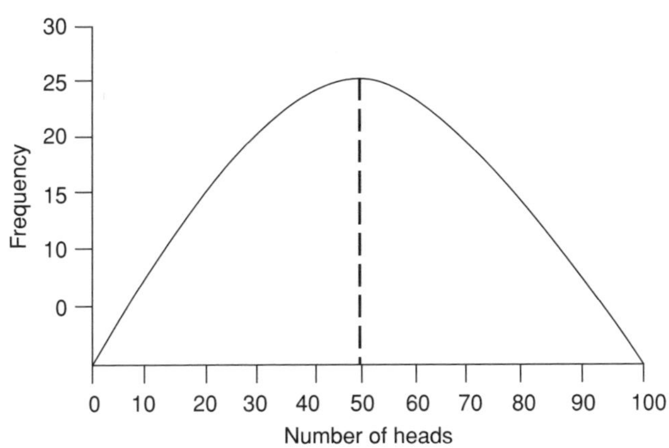

Figure 1.3 Normal distribution curve

Thus, in effect, Six Sigma is a theoretical statistical measurement allowing the assessment of the quality of products and services to a position where there exist practically zero defects for any product or process in an organisation.

The key success factors differentiating Six Sigma from TQM are:

- Emphasis on statistical science and measurement.
- Structured training plans at different levels (Champion, Master Black Belt, Black Belt and Green Belt).
- Project-focused approach with a single set of problem-solving techniques such as DMAIC (Define, Measure, Analyse, Improve, Control).
- Reinforcement of Juran's tenets such as top management leadership, continuous education and annual savings plan.
- Effects quantified in tangible savings (as opposed to TQM, where it was often said that we can't measure the benefits, but if we didn't have TQM who knows what losses might have occurred).

This last point concerning the quantification of tangible savings is a major selling point for Six Sigma.

It is usually possible to measure the cost of poor quality (COPQ) with the sigma level at which the process consistently performs. The COPQ if the performance level is at six sigma will be less than 1% of cost of sales, whereas at three sigma (three sigma is regarded by many organisations as a very acceptable level of process quality) the corresponding COPQ will range from 25% to 30% of cost of sales. These figures are explained/justified later in this book.

1.7 THE STRUCTURED APPROACH OF SIX SIGMA

Following the rigorous application of Six Sigma in many organisations, including Motorolla, Allied Signal, General Electric, Bombardier, ABB, American Express, Wipro, GSK and others, a proven structured approach has emerged for product and process improvement. This structured and hierarchical process is shown below:

EXECUTIVE LEADERS/SPONSORS

↓

MASTER BLACK BELTS
(also known as Champions)
Owners of critical projects

↓

BLACK BELTS

(also known as Experts)
Change agents assigned to Masters
to tackle critical projects

↓

GREEN BELTS

(also known as Agent Advocates)
Grass root support to implement changes

The top management must have a total commitment to the implementation of Six Sigma and accomplish the following tasks:

1. Establish a Six Sigma leadership team.
2. Develop and roll out a deployment plan for the training of Master Black Belts, Black Belts and Green Belts.
3. Assign Master Black Belts to identify and own critical projects related to key business issues.
4. Provide support for Black Belts to make breakthrough improvements in critical projects.
5. Encourage Green Belts to identify and implement 'just do it' projects.
6. Set aggressive Six Sigma targets. This is done by the Six Sigma leadership team.
7. Continuously evaluate the Six Sigma implementation and deployment programmes and make changes if necessary.

1.7.1 Six Sigma Deployment

A critical piece of successful Six Sigma experience is the Six Sigma deployment plan. A typical plan includes four parts:

1. Business alignment planning.
2. First wave of Black Belt training.
3. Second wave (and subsequent waves as required) of Black and Green Belt training.
4. Infrastructure development to deliver results and sustain culture.

Business alignment planning assures that Six Sigma projects align with business strategy and drive results. Preliminary project selection criteria start with projects having potential substantial savings (1 million US dollars or more). During the leadership education programme (which usually lasts from five days), Master Black Belts are selected and Masters are assigned to key business issues and projects. Intensive Master Black Belt instruction takes three to four weeks. Concurrently while the Masters are being trained, the selection process for those to be qualified as Black Belts is carried out. Master Black Belts (MBBs) have a thorough understanding of the improvement process, DMAIC, Six Sigma-associated statistics and change management.

An MBB is competent to lead and manage significant end-to-end projects. They are also capable of coaching Green Belts.

Black Belt training is a combination of formal classroom education and onsite project work. Teaching is spread over four to five months, and includes four or five weeks of classroom work with the balance being onsite project experience (see Table 1.2).

Green Belts (GB) have an awareness of Six Sigma principles and in particular apply the DMAIC cycle to work with and support Black Belts in end-to-end projects. Green Belt training is usually spread over six or perhaps seven weeks, including five days of formal classroom work.

External consultants and experienced trainers are usually needed to manage and train the first deployment wave. For the second and third waves of the deployment plan, in-house Master Black Belts take charge of both the training and the management of specific projects.

Table 1.2 Sample of Black Belt training

Week 1	Week 2	Week 3	Week 4
Define and measure	Analyse	Improve and control	Advanced statistical techniques
Variation	Advanced statistics	Generating improvement	Design of experiments
Data collection			Process capability
Basic statistics		Change management	Quality function deployment
Cost of poor quality (COPQ)			Multivariance analysis
Define and measure tools	Analyse and SPC tools	Improve and control tools	Design for Six Sigma

1.7.2 Certification of Black Belts and Master Black Belts

The certification authority is the Six Sigma Academy. They have set standards and guidelines as follows.

Black Belts are awarded a certificate after:

- Completing three weeks of formal classroom Black Belt training.
- Completing one week of change management and project management training.
- Completion, as a team leader, of an end-to-end $1 million Black Belt project.
- Demonstrated Six Sigma commitment evidenced by mentoring Green Belts and preparation of Green Belt training material.

Master Black Belts are awarded a Masters certificate after:

- Black Belt certification.
- Completion of five end-to-end Black Belt projects as team leader.
- Delivery of three Green Belt training sessions.
- Demonstrated commitment to the Six Sigma philosophy through mentoring Black Belt and/or Green Belt projects.
- Designing Black Belt training materials.

1.8 WHAT IS LEAN SIGMA?

Lean Sigma (also known as LSS) is the combination of two world-class approaches (Six Sigma and Lean Manufacturing) to organisational performance improvement. You have been introduced to Six Sigma in the preceding section. Let us explain Lean Manufacturing and then show how these two approaches are combined in Lean Sigma.

The term 'lean manufacturing' was first employed by Womack and Jones in 1998 as a description of the way manufacturing was carried out at Toyota. Toyota called their manufacturing process the Toyota Production System (TPS). Ohno (1988), who is said to be the founder of TPS, noted that the basis of this method is the elimination of waste. In order to achieve this, two pillars are used:

- just in time;
- automation with a human touch.

'Just in time' in a production means that the right parts reach the assembly line at the time they are needed and only in the amount required. The basic tool for achieving just in time is the Kanban system. 'Automation with a human touch' goes back to Sakichi Toyoda, the founder of Toyota, who invented the automatic loom. This loom would automatically stop if a thread broke, thereby preventing any work being carried out after the defect had occurred. According to Ohno (1988), an important part of TPS is the ability to prevent mistakes from happening again. This is achieved by solving the cause of the problem rather than the actual mistake. Examples of tools for problem solving and finding root causes of difficulties are: Genshi Genbutso, which means going to the source to directly observe the problem; cause/effect diagrams (also known as fish-bone diagrams); and the '5 whys', or asking questions five times.

Ohno (1988) identified seven types of waste or 'mudas':

1. Waste of overproduction.
2. Waste of time on hand (waiting).
3. Waste in transportation.
4. Waste of processing itself (processing too much).
5. Waste of stock on hand (inventory).
6. Waste of movement.
7. Waste of making defective products.

According to Ohno, the biggest source of waste is overproduction as it leads to many of the other types of squandering and misuse.

Liker (2004) has written several books about Toyota and he describes 14 management principles garnered from the company. Liker writes in his book that lean is not just about implementing tools such as 5S and just in time. It also concerns applying a complete system of Lean Thinking and Culture. This view is supported by Womack and Jones (2003) in a concept for Lean Thinking, as 'Lean Thinking is lean because it provides a way to do more and more with less and less'. Their five principles of Lean Thinking are Value, Value Stream, Flow, Pull and Perfection.

However, the application of Lean Thinking has moved with the times and the experience of organisations in both the manufacturing and service sectors. Basu and Wright (2008) have extended Lean Thinking to supply-chain management. The competition for gaining and retaining customers and market share is between supply chains rather than other functions of companies. A supply chain therefore has to be lean, with four inter-related key characteristics or objectives:

1. Elimination of waste.
2. Smooth operation flow.
3. High level of efficiency.
4. Quality assurance.

1.8.1 Elimination of Waste

The lean methodology as laid out by Womack and Jones (1998) is sharply focused on the identification and elimination of 'mudas' or waste. Indeed, their first two principles (i.e. Value and Value Stream) are centred around the elimination of waste. Their motto has been 'banish waste and create wealth in your organisation'. It starts with value stream mapping to identify value and then pinpoints waste with the process mapping of valued processes, followed by systematically eliminating them. This emphasis on waste abolition has probably made lean synonymous with the absence of waste. Waste reduction is often a good place to start in the overall effort to create a lean supply chain because it can often be achieved with little or no capital investment.

1.8.2 Smooth Operation Flow

The well-publicised just-in-time approach (JIT) is a key driver of the lean supply chain and, as we have indicated earlier, it requires that materials and products flow 'like water' from the supplier through the production process on to the customer. The capacity bottlenecks are eliminated, the process times of work stations are balanced, and there are few buffer inventories between operations. Smooth operation flow requires the applications of appropriate approaches. Three of the most frequently applied methods are:

- Cellular manufacturing.
- Kanban pull system.
- Theory of constraints.

In the cellular manufacturing concept, the traditional batch production area is transformed into flow line layouts so that ideally a single piece flows through the line at any time. In practice, an optimum batch size is calculated starting with the most critical work centres and the largest inventory carrying costs. Action is taken for improvement both at these work centres and concerning methods that have greatest impact on the throughput, customer satisfaction, operating cost and inventory carrying charges. Secondly, Kanban (literally meaning 'card') is a way of pulling parts and products through the manufacturing or logistics sequence as needed. It is therefore sometimes referred to as the 'pull system'. Finally, 'the theory of constraints' is a management philosophy developed by E.M. Goldratt (1992). It enables the managers of a system to achieve more of the goal that system is designed to produce. However, the concept or the objective is not new. Nonetheless, in service operations where it is often difficult to quantify the capacity constraint, TOC could be very useful.

1.8.3 High Level of Efficiency

The more popular concepts of Lean Operations tend to be the principles of mudas, flow and the pull system. However, a preliminary analysis of all these methods, as we have described earlier, highlights the fact that all assume sufficient machine availability exists as a pre-requisite. In our experience, for many companies attempting a lean transformation, this assumption is just not true. Machine availability depends on maximising the machine up time by eliminating the root causes of down time. The ratio of up time and planned operation time is the efficiency of the operation. Therefore, in order to make lean concepts work, it is vital that the pre-condition of running the operations at a high level of efficiency should be met.

1.8.4 Quality Assurance

Womack and Jones (1998) propose 'Perfection' as the fifth Lean principle. According to this tenet, a lean manufacturer sets his or her targets for perfection in an incremental (Kaizen) path. The idea of TQM is also to systematically and continuously remove the root causes of poor quality from the production processes so that the organisation as a whole – and its products – are kept moving towards perfection. This relentless pursuit of the faultless has to be a key attitude of an organisation that is 'going for lean'.

Table 1.3 DMAIC and SCORE

Six Sigma (DMAIC)	Lean Kaizen (SCORE)
Define	Select
Measure	Clarify
Analyse	Organise
Improve	Run
Control	Evaluate

1.9 MORE ON LEAN SIGMA

If the objective of Six Sigma is the reduction of variation, then Lean Sigma aims to accomplish the mission of the organisation better, faster and cheaper. To put it another way, Lean Sigma combines the focus on efficiency by Lean and the emphasis on quality by Six Sigma. There is a trade-off between quality (better), faster (delivery time) and cheaper (cost). The key is to leverage Lean Sigma to enable the organisation to get better, faster and cheaper – all at the same time. In general, Six Sigma is used to reduce defects and errors, thus making products or processes better. However, Six Sigma also reduces waste and streamlines processes like Lean. In fact, although Lean is primarily used to remove waste, in doing so it also reduces rework. When you reduce rework you also reduce defects. Therefore, it is more sensible to combine the two methodologies of Lean and Six Sigma to make one approach of Lean Sigma. For example, consider a problem of lead time which varies between 20 and 12 weeks with an average of 16 weeks. With a Six Sigma approach it now fluctuates between 18 and 17 weeks with a mean value of 16 weeks. When Lean Sigma is applied the lead time is reduced to 10 weeks, varying between 11 and 9 weeks.

There appear to be two approaches to Lean Sigma (De Carlo, 2007). One is to follow the DMAIC methodology of Six Sigma, to focus on the business objective of value creation and to apply additional Lean tools such as value stream mapping and rapid changeover. The second approach is centred on Kaizen Event (SCORE) rather than DMAIC. It is argued that the DMAIC process often takes longer when root causes are not known. On the other hand, you may employ a Kaizen event when the root causes are known (e.g. for cycle time reduction). This may follow a war-room-like environment for a short period (also called the Kaizen Blitz). However, a drawback of this process is that it may not bring the rigour of data-driven measurement and analysis of DMAIC. As shown in Table 1.3, although there are similarities between DMAIC and SCORE, the former is more related to project management and the latter is more linked to industrial engineering.

1.10 WHY FIT SIGMA?

We live in a competitive world. The pace of change is increasing, and businesses and national economies are continuously being disrupted. In recent times the biggest external factor has been the collapse of banking and the finance market, causing job losses and a sharp decline of gross domestic products. We have also experienced the impact of the Internet and e-business. The spectacular rise in 1999 and fall in 2000 of so many so-called dot.com companies showed that without substance, businesses are not sustainable. When the huge bubble bursts it is the innocent consumers who will feel the effect. Added to this, we need to factor in large-scale mergers, and acquisitions are continuously taking place in all sectors of manufacturing and service industries. Globalisation and the economic growth, particularly in China, are reshaping the economic power balance and consequent business strategies. We can conclude that change is here to stay; and it often comes quickly from unexpected quarters. The challenge for all businesses is to find the benefits of change with an appropriate value-adding change programme.

The failed dot.com companies of the last decade and the current problems of the banking industry have demonstrated that their early apparent success was not sustainable. There are similar stories of unsustainable improvements in traditional businesses in the 'old economy' (prior to e-business). In spite of the demonstrated benefits of many improvement techniques such as TQM, Business Process Re-engineering and Six Sigma/Lean Sigma, many attempts to implement and uphold improvements have fizzled out, not with a bang but with a whimper. What is more puzzling is that some companies (such as Motorola and Ford), who successfully implemented a quality initiative (such as Six Sigma), have subsequently experienced overall drops in business performance resulting in lay-offs and lower employee morale. Therefore, we need a sustainable continuous improvement programme.

The well-publicised results of Six Sigma were first by Motorola, General Electric and Allied Signal and then by Ford and Dow Chemical. This created an image of Six Sigma as a process primarily for large manufacturing multinationals who can sanction a huge budget for consultants to train hundreds of Black Belts. In addition to the apparent misconceptions of high start-up cost and only being suitable for manufacturing, an aversion to sophisticated statistical techniques also discouraged the service sector and small and medium enterprises to apply Six Sigma. Therefore, there is a need for a 'fit for the purpose' improvement programme suitable for all types or sizes of organisation.

As described in the following chapters, the holistic and appropriate methodology of FIT SIGMA aims to address these gaps.

1.11 SUMMARY

Quality management has evolved over the years through various stages, from inspection to control to assurance to TQM. The new waves of Six Sigma, Lean Processes and FIT SIGMA™ are embedded in the holistic programmes of operational excellence. Our belief is that quality is not a new or separate discipline, but rather that quality pervades all management actions. Our philosophy is that quality is too important to be left to the managers and that quality is everybody's concern, not only in the organisation, but also customers, suppliers and any other stakeholder.

Quality has two main aspects. It can be measured from the customer's perspective – customer satisfaction – and it can be viewed from the perspective of the efficient use of resources. These two seemingly separate objectives are in fact inseparable when quality is considered. An organisation that wishes to compete in the global market must be efficient and provide a high level of customer satisfaction. No organisation will be able to afford to provide world class service unless its use of resources is efficient and non-value-adding activities have been minimised, and no organisation can afford to be not world class.

With the above logic, the application of a Six Sigma or Lean Sigma programme should address both customer service and resource utilisation in a cost-effective way. Quality has three dimensions – product quality, process quality and organisation quality (Basu, 2004). The results should also be sustainable and underpinned by people-related organisation quality. As explained in the following chapters, this is wherein lies the value of FIT SIGMA.

2
More about FIT SIGMA

2.1 INTRODUCTION

The success of Six Sigma and Lean Sigma cannot be faulted. The rigorous Six Sigma process combined with the speed and agility of Lean Sigma has produced definitive solutions for better, faster and cheaper business processes. Through the systematic identification and eradication of non-value-added activities an optimum value flow is achieved, cycle times are reduced and defects eliminated. However, business managers do understand the grey areas of distinction between different quality initiatives and justifiably are expressing concerns, including raising the question 'how do we sustain the results?'

A survey by Basu[1] has shown that there are considerable barriers to achieving and sustaining results in quality initiatives. These are illustrated in Figure 2.1.

The biggest obstacle appears to be the packaged approach of the programme, causing a paucity of customised local solutions. Furthermore, due to the 'top-down' directive, middle managers are often not 'on board'. The initiative is not owned by employees. We can identify additional and complementary areas of concern, including:

- 'Some star performers of Six Sigma have shown poor business results' (e.g. site closures by Motorola).
- 'Incomplete initiatives' (e.g. Marconi abandoned Six Sigma during the economic downturn of 2001).
- 'Change of management and loss of sponsors' (e.g. the decline of Six Sigma at Allied Signal after the departure of Larry Bossidy).
- 'External push by high powered consultants' (e.g. the dominance of the Air Academy consortium in the GSK programme).
- 'Excellent early results not sustained' (e.g. Ratheon relaunched Lean Sigma after a drop in performance).
- 'High start-up costs impede small and medium sized enterprises' (the initial training start-up cost for Six Sigma is reported to be over $1 million).
- 'Still regarded as tools for manufacturing' (in spite of the success of Six Sigma in GE Capital).

The dramatic bottom line results and extensive training deployment of Six Sigma and Lean Sigma must be sustained with additional features in order to secure the long-term competitive advantage of a company. If Lean Sigma provides agility and efficiency, then measures must be in place to develop a sustainable fitness. The process to do just that is FIT SIGMA.[2] In addition, the control of variation from the mean in the Six Sigma process (σ) is transformed to company-wide integration in the FIT SIGMA process (Σ). 'FIT SIGMA' is therefore synonymous with 'FIT Σ'. Furthermore, the philosophy of FIT Σ should ensure that it is

[1] Unpublished GSK survey, January 2001.

[2] FIT SIGMA is a registered trademark of Performance Excellence Ltd.

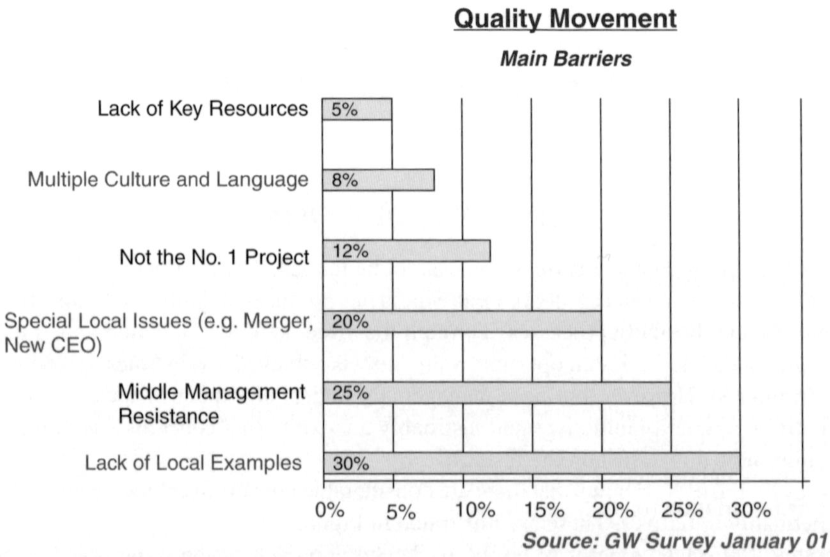

Figure 2.1 Main barriers to quality movement

indeed fit for all organisations – whether large or small, manufacturing or service. In this book both FIT SIGMA and FIT Σ should be treated as identical terms.

FIT Σ is a solution for *sustainable* excellence in *all operations*. It is a quality process beyond Six Sigma. The fundamentals of FIT Σ are underpinned by the three cornerstones as shown in Figure 2.2.

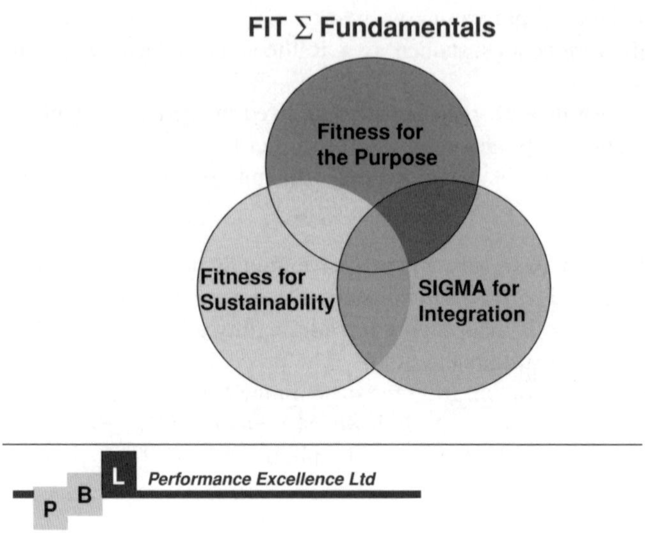

Figure 2.2 FIT Σ fundamentals

Taken individually, the main components of these cornerstones are not new, but they do constitute proven processes. However, the combination of these components is both novel and unique, to create the total process called FIT Σ. These elements are as follows.

FITNESS FOR THE PURPOSE

- Initial assessment.
- All functions.
- Any size of organisation.
- Aligning with business strategy.

SIGMA (Σ) FOR IMPROVEMENT AND INTEGRATION

- Appropriate Six Sigma tools (see Chapter 4).
- Appropriate DMAIC methodology (see Chapter 3).
- Learning Deployment.
- Project plan and delivery.
- Shift from variation (σ) to integration (Σ).

FITNESS FOR SUSTAINABILITY

- Performance management.
- Self-assessment and certification.
- Senior management review (S&OP).
- Knowledge management.

2.2 FITNESS FOR THE PURPOSE

It was Joseph Juran who coined the phrase 'fitness for the purpose', relating to the basic requirements for quality. In the context of FIT Σ, 'fitness for the purpose' has wider implications. Here, 'fitness' means that the FIT Σ methodology is tailored to 'fit' all types of operations (whether manufacturing, service or transport) as well as all sizes of organisation (whether a multi-billion dollar global operation or a small local enterprise). The customisation of the improvement programme to identify the right fit appropriate to the type and size of operation is determined by a formalised initial assessment process.

Treacy and Wiersema (1993) described three value disciplines in a business strategy:

- Operational excellence.
- Customer intimacy.
- Product leadership.

They suggested that in a business context, organisations must select and excel at one of these value disciplines as a core operating model, while remaining adept at the other two. Regardless of an organisation's specific core competence or the sector in which it operates (e.g. manufacturing, retail, service or technology), the value discipline shapes the choices made by managers on a day-to-day basis. Applying the same logic in FIT SIGMA, each organisation must align the objectives and scope of the programme with the value disciplines of the business strategy.

It is advantageous, although not essential, to apply the initial assessment process based on a set of questions which could be applied later for the periodic self-assessment or certification process. This certification process, which will be described in more detail later, may be adapted either from international quality awards (such as the European Foundation of Quality Management or America's Malcolm Baldridge Quality Award) or a holistic, published checklist (such as Basu and Wright's Total Solutions 200 Questions), or even the company's own inventory. Regardless of the scheme, it is essential that the checklist is customised and the key players of the company believe in it. As part of the Six Sigma or Lean Sigma programme, a 'base line analysis' is carried out to identify areas of improvement, but this is performed at a later stage after committing the company to a rigorous deployment programme. The initial assessment is similar to that of a 'base line analysis', but it is carried out right at the onset before the start of the deployment plan.

We believe that there are substantial benefits to be gained from designing your own assessment process rather than following a consulting firm's standards. However, it is imperative that the company is aware of the requirements and criteria of the assessment. The initial evaluation requires a good understanding of one's own processes, while an objective training to international standards is necessary for the certification. In both cases, the assessor is a trained expert.

Case example: Six Sigma in a small and medium enterprise (SME)

The Solectron factory in Ostersund, Sweden, where AXE switchboards are manufactured, employs approximately 1000 people. The site was formerly part of the Ericsson Network of core products AB. Solectron as an independent company was experiencing tough competition even at the crest of the 'telecom boom'. With the downturn in the market from 2000, the competition became increasingly fierce. The management were toying with the idea of launching a Six Sigma initiative, but their initial enquiry revealed that they would be set back by at least $1 million if they began a formal Black Belt training programme with the Six Sigma Academy at Scottsdale, Arizona.

Ericsson, the parent company of previous years, had already embarked upon a Six Sigma initiative. The Black Belt training programme was also in full swing. Solectron decided to send a promising manager to a Black Belt training course via the Ericsson deployment plan. The young manager duly returned to Ostersund with great enthusiasm and applied a preliminary 'base line analysis' rooted in a simple checklist. The results were then presented to the management and a customised programme was drafted. The training programme was extensive, but Solectron relied on the Black Belts from Ericsson and also retained the same consultants as and when required.

Ten members of the top management team attended a one-day course on Six Sigma, 14 people were trained as 'Black Belts' on a seven-month part-time programme and 20 more attended a two-day course. Six Sigma applications at this factory saved US$0.5 million during the first year of the project. This amounted to about $500 per employee, but was actually closer to a huge $36,000 per employee trained in Six Sigma methods. A modest start in terms of savings per Black Belt perhaps, but the investment was also a fraction of a 'pure play' Six Sigma initiative. More significantly, this customised approach enabled Solectron to have a launch pad to gain a much needed competitive advantage.

Solectron applied the FIT Σ methodology – 'Fitness for the purpose' – albeit not consciously under any label.

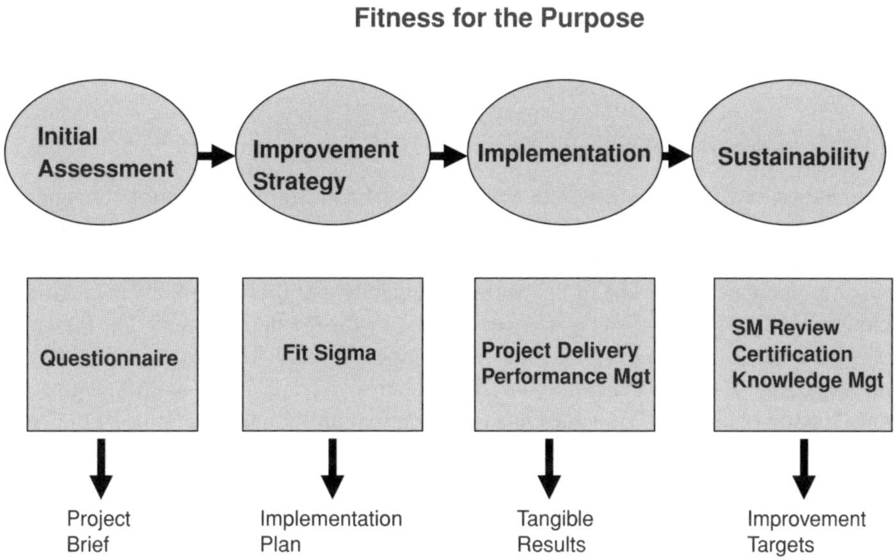

Figure 2.3 Fitness for the purpose

Many organisations have dived into a Six Sigma programme without going through the earlier crucial stages of identifying the real requirements. Arguably, a significant number of companies that initiated a Six Sigma programme did so because they felt threatened in terms of their very survival or became victims of management fads. The 'GE factor' was too strong to ignore. But we should not interpret this 'GE factor' as a licence to copy blindly General Electric, but rather an exhortation to 'get effective' by applying the FIT Σ methodology. The starting point is an initial assessment as illustrated in Figure 2.3.

The initial assessment process can benefit from the application of a rating scale to each criterion or question on a ladder of say 1 to 5, with 1 being marked as 'poor' and 5 ranked as 'excellent'. A spider diagram (see Figure 2.15) can be constructed from the scores of each criterion to highlight the gaps. The assessment process is described in more detail later in Section 4.3.

The 'fitness for the purpose' philosophy also applies to the type of business, whether it is manufacturing operations in a factory or a service procedure in an office. The success of GE Capital after the application of Six Sigma is well documented. However, Jack Welch has often been quoted as the success factor rather than the application of the principles of Six Sigma and, after all, GE Capital is a huge operation of billions of dollars.

2.3 SIGMA (Σ) FOR IMPROVEMENT AND INTEGRATION

The key cornerstone of FIT Σ methodology is 'Sigma (Σ) for improvement and integration'. The 'improvement' aspect is essentially based upon the proven tools and processes of Six Sigma/Lean Sigma initiatives. While the contents of Six Sigma or Lean Sigma approaches vary according to a company, a consultant or an author, the common features with FIT Σ are:

- Appropriate Six Sigma tools.
- Appropriate DMAIC methodology.

- Learning deployment.
- Project plan and delivery.
- Shift from variation (σ) to integration (Σ).

2.3.1 The Appropriate Six Sigma Tools

Under detailed analysis, it is true that Six Sigma or Lean Sigma tools are not original. For example, the focus on variation is historically known as the control chart of Deming and Shewhart. Design of experiments can be linked to Taguchi's methods. The proactive use of Pareto and 'fishbone' diagrams in Six Sigma is laudable, but these are really 'old hat', being the clothing of TQM and tools originally developed by Pareto and Ishikawa. The flow process chart of Lean Sigma is also a classic industrial engineering tool. Therefore we do not propose to introduce any so-called 'FIT Σ tools', but rather to refer to them as 'Six Sigma tools'. Thomas Edison once said, 'Your idea has to be original only in its adaptation to the problem you are currently working on'. The adaptation of the existing apparatus is the 'appropriate Six Sigma' tools. These include those listed in Table 2.1.

Some of the tools outlined above have already been described, and further details can be found in Appendix I. We recommend that the learning deployment of FIT Σ should ensure both the understanding and application of these 'appropriate tools'.

Chapter 4 describes suitable FIT SIGMA tools.

2.3.2 Appropriate DMAIC Methodology

The basic framework of DMAIC cannot be doubted and it is also a core requirement of FIT SIGMA. However, the time and cost of using DMAIC should be weighed against the complexity and potential benefits of each project. Depending on the requirements of an individual undertaking, three levels of methodology are suggested:

- DMAIC Full.
- DMAIC Lite.
- KAIZEN Event.

Table 2.1 Six Sigma tools

Basic tools
- Pareto diagram
- Flow process chart
- UCL/LCL control chart
- Cause and effect diagram
- Input–process–output (IPO) diagram
- Brainstorming
- Scatter diagram
- Histogram
- The seven wastes
- The five Ss
Advanced tools
- Failure mode and effect analysis (FMEA)
- Design of experiments (DOE)
- Design for Six Sigma (DFSS)

Table 2.2 Deployment plan outline

Programme	Target audience	Duration	Approximate number
Leadership education	Senior management	2 days	3–5% of employees
Expert training (Black Belt)	Senior and middle management	4–6 weeks (in waves over 6 months)	1% of employees
Advocate training (Green Belt)	Supervisors and functional staff	1 week	10% of employees
Appreciation and cultural education	All employees	2 × half days	All

Chapter 3 describes the above levels of methodology and the tips for their appropriate selection.

2.3.3 Learning Deployment

In order to use tools 'appropriate for Six Sigma' to achieve longer-term benefits, it is essential that an extensive learning deployment programme is dedicated to the education and training of employees at all levels. The learning deployment plan should be formulated after the 'initial assessment' and details will vary according to the 'quality level' and size of the organisation. We recommend that the proven paths of previous Six Sigma and Lean Sigma programmes should be treated with respect and a deployment plan can then be built around the outline shown in Table 2.2.

Through a rigorous training deployment programme, Six Sigma has contributed to the creation of a people infrastructure within an organisation to roll out a comprehensive programme. The issue is not who should be trained, but rather: who should do the training? The original source of Six Sigma education was the Six Sigma Academy, founded in 1994 in Scottsdale, Arizona. It is run by former Motorola experts Mikel Harry and Richard Schroeder. Their fees have been reported to start at $1 million per corporate client. Although there are many capable consultancy firms offering instruction, such tutoring costs are still running on an average threshold of $40,000 per 'Black Belt'.[3] Thus, the initial start-up and training expenses have prevented many small and medium sized companies from embracing a Six Sigma programme.

It is essential that high-quality input is provided to the training programme. This is usually available from specialist external advisers. At the same time, a 'turn key' consultancy support is not only expensive but also contains the risk that 'when consultants leave, expertise leaves as well'. In the FIT Σ learning deployment programme, we recommend two options:

Option 1

- Retain consultants for, say, three months.
- Run part of Leadership Education and Expert Training programmes.
- Develop, with the assistance of consultants, own trainers and experts to complete the remaining waves of Leadership and Expert Training.
- Coach advocates (Green Belts) by own experts.
- Cultural education by line managers.

[3] Unpublished GSK survey, January 2001.

Option 2

- Deploy consultants for one top-level Leadership Education Workshop.
- Prepare a small team of experts (two to five people) as trainers and develop a deployment plan with the assistance of consultants.
- Roll out the deployment plan with own experts.
- Ensure that consultants are available if required.
- Train advocates (Green Belts) by own experts.
- Cultural education by line managers.

In General Electric, known to be the 'cathedral' of Six Sigma, the Six Sigma programme has been supported globally by a corporate team (known as CLOE – Centre of Learning and Operational Excellence), based at Stanford, Connecticut. Regardless of the type or size of operations, the development of a firm's own training capability is the foundation of sustainable performance.

Case example: Six Sigma training deployment in Noranda Inc.

Noranda Inc. is a leading international mining and metals company for copper, zinc, magnesium, aluminium and the recycling of metal. With its headquarters in Toronto, the company employs 17,000 people around the world and its annual turnover in 2000 was $6.5 billion.

In August 1999, the Board of Noranda decided to embark upon a global Six Sigma project with an initial savings target of $100 million in 2000. There were, however, some specific challenges to overcome. The company business is in a traditional industry with long-serving trade. Furthermore, Noranda is a 'decentralised' organisation with multiple cultures and languages. Senior executives studied the experiences of other companies (GE, Allied Signal, Dupont, Bombardier and Alcoa) and invited the Six Sigma Academy from Arizona to launch the training deployment programme.

The Six Sigma structure at Noranda focused on the training of the following levels:

Deployment and Project Champions
Master Black Belts
Black Belts
Business Analysts ad Validates
Process Owners
Green Belts

The Six Sigma Academy was intensely involved for the first three months of the programme and then Noranda started its own education and training. The training accomplishments in 2000 were impressive:

All 84 top executives followed a two-day workshop.
90 Black Belts were certified.
31 Champions were trained.
There were 17 days of Master Black Belt training.
3000 days of Green Belt training.
More than 3500 days of training in 2000 – and this has continued.

The learning deployment to educate and develop your own experts or Black Belts provides a successful balance between a well-measured job structuring by a central team of industrial engineers, and self-managed work teams. Over the years the principles of industrial engineering and Taylorism became corrupted to the extent where time and motion study people found the best method and then imposed that best method upon the worker. Such external control impeded team work and created the tedium of repetitive operations. This was followed by the 'quality circles' (mainly in Japan) and self-managed teams (mainly in Scandinavia). The failure of the experimental Volvo factory at Uddevalle in the early 1990s was a wake-up call to realise that planners and team leaders should be trained in analytical tools. The experience of both GM–Toyota joint ventures in California and the Ford–VW factory in Portugal demonstrated that group performance can be improved by training the teams in industrial engineering principles. The collaboration of expert (or Black Belt) training and the team comprising advocates (or Green Belts) provides a balance of empowerment, motivation and measured efficiency.

2.3.4 Project Plan and Delivery

The success of any project is underpinned by management commitment, organisation, resources and formal reviews. A FIT Σ programme is no exception to the basic rules of project management. The process logic of a FIT Σ programme is shown in Figure 2.4, where the positions of a project plan and delivery have been highlighted.

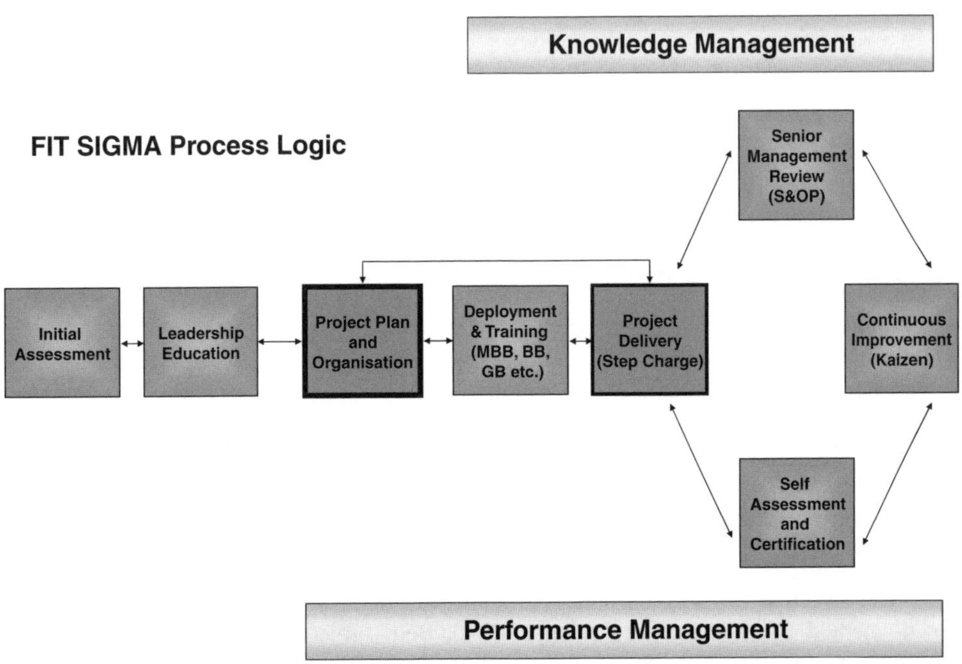

Figure 2.4 FIT Σ process logic

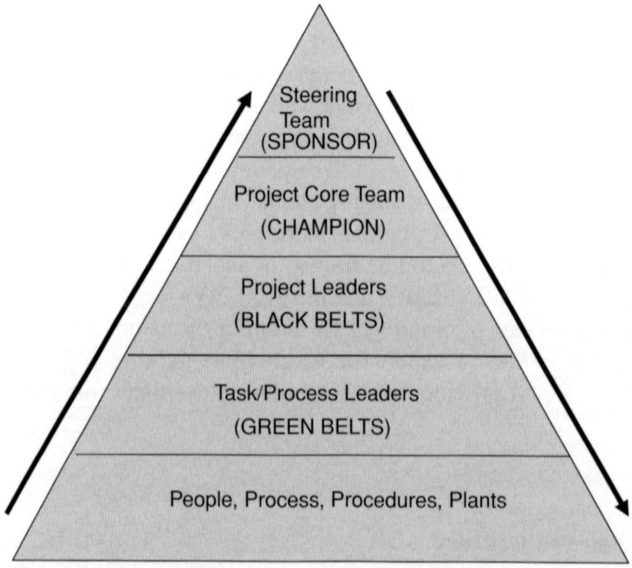

Figure 2.5 Project organisation: a two-way communication process

The structure of a project organisation varies according to the operations and culture, but it must comprise some essential requirements:

- A sponsor or 'torch bearer' should be at the highest level of the organisation.
- Project team leaders should be multi-functional and 'Black Belt' trained.
- Two-way communication – both top-down and bottom-up.
- RACI (responsibility, accountability, consulting and information) roles should be defined clearly.

Figure 2.5 shows an example of a typical project organisation for a FIT Σ programme.

We recommend at an early stage, following the initial assessment, that a project brief or project charter is prepared to define clearly:

- project organisation;
- time plan;
- learning deployment;
- project selection criteria;
- key deliverables and benefits.

The project selection criteria cover two broad categories of ventures within the FIT Σ programme:

1. Large projects (managed by Black Belts).
2. Small 'just do it' assignments.

Project selection can rely on both the 'top-down' and 'bottom-up' approach. The 'top-down' method usually relates to a large 'Black Belt' project and considers a company's major business issues and performance objectives. Teams identify processes, CTQ (critical to quality) characteristics, process base lining and opportunities for improvement. This tactic has the advantage of aligning FIT Σ projects with strategic, corporate objectives. The 'bottom-up' approach can apply to both large and 'just do it' schemes.

A rule of thumb for large Six Sigma projects was to attain savings of $1million per year per Black Belt, deriving from between four and six projects each year. A 'just do it' task usually does not incur a significant investment and varies from producing $5000 to $100,000 savings annually.

Adapted from Oakland (2003), we summarise five key sources for identifying savings projects:

1. DPMO (defects per million opportunities) – The number of defects per unit divided by the number of opportunities for defects multiplied by 1,000,000. This number can be converted into Sigma Value.
2. COGS (cost of goods sold) – The variance is fixed and variable costs.
3. COPQ (cost of poor quality) – Comprising internal failure costs, external failure costs, appraisal costs, prevention costs and lost opportunity costs.
4. Capacity – Overall equipment effectiveness; the number of good units a process is able to produce in a given period of time.
5. Cycle time – The length of time it takes to produce a unit of product or service.

It is important that, at the early stage of the programme, both large and 'just do it' projects are straightforward and manageable. These types of venture are often referred to as 'low-hanging fruits' – where improvements can be achieved easily by 'basic tools' (e.g. fishbone diagrams, flow process charts, histograms, etc.).

A simple selection process has been developed by Johnson and Johnson at their Wappertal plan in Germany. A rating of projection selection criteria related to savings and ease of implementation was applied. The projects were selected after plotting the scores on a graph as shown in Figure 2.6.

However, FIT Σ is far from completing savings projects. Over time, undertakings are selected and implemented to strengthen the company's knowledge base, stabilise processes and procedures, and expand the cross-functional boundaries.

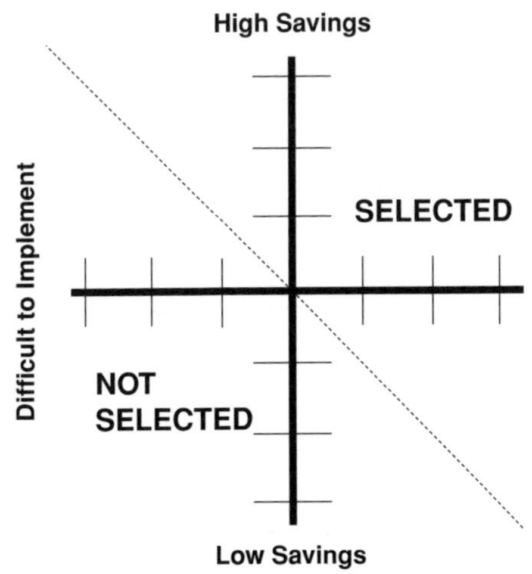

Figure 2.6 Project selection

Case example: A Six Sigma project in the Dow Chemical Company

Film Tec Corp. is a subsidiary of the Dow Chemical Company in Minneapolis and a manufacturer of water purification membranes. The quality of a membrane is determined by two criteria: flux, or the rate of water that the membrane lets through, and how much impurity is removed from the water. At Film Tec, membrane elements were tested prior to shipping to customers. The quality of membranes the customers received was protected but the speed with which they were serviced suffered. Rejected products were costing Film Tec approximately $500,000 a year.

With a focus on customer needs, the Six Sigma project team required a strategic shift in participation by all employees to analyse systematically the internal manufacturing procedures. A key variable identified for improvement was the inconsistencies in the concentration of a chemical component used in the manufacturing process. The problem stemmed from the interruption to batch feeding the chemical into the manufacturing process. To reduce the variation, an inexpensive reservoir was added to feed the chemical while containers were exchanged. Additionally, a level transmitter with an alarm was installed to alert the operators to the low level of this chemical.

The improvements have been significant. The reject rate was reduced from 14.5% down to 2.2%. In addition to these savings for Film Tec, membranes are available to customers faster than before.

2.3.5 Shift from Variation (σ) to Integration (Σ)

Our FIT Σ process fully accepts the importance of variation reduction. The risk of an improvement process based upon average values alone has been incontrovertibly proven. Likewise, there is an abundance of real-life examples where added values of lower variation or 'span' are well established. It is essential to concentrate on the variation control of sub-systems and individual processes. However, the Six Sigma theme of variation control has often caused the focusing of an improvement plan in a relatively narrow sector or department.

Bob Glavin, former CEO of Motorola, has stated that the lack of an initial Six Sigma initiative in non-manufacturing areas was a mistake that cost Motorola $5 billion over a four-year period.

www.ariacad.com

The success of Six Sigma within General Electric was further enhanced by moving from a 'quality focus' to a 'business focus' and extending the initiative to its financial services area (e.g. GE Capital).

It is indicative, though not conclusive, that maximum benefit will be obtained from Six Sigma by integrating it with other proven continuous improvement initiatives and extending the programme to encompass the total business. When that happens, then Six Sigma embraces the FIT Σ philosophy of integration. The shift becomes complete from a small σ (standard deviation) to a capital Σ (summation or integration).

Figure 2.7 illustrates a mapping of the shift from Six Sigma to FIT Σ.

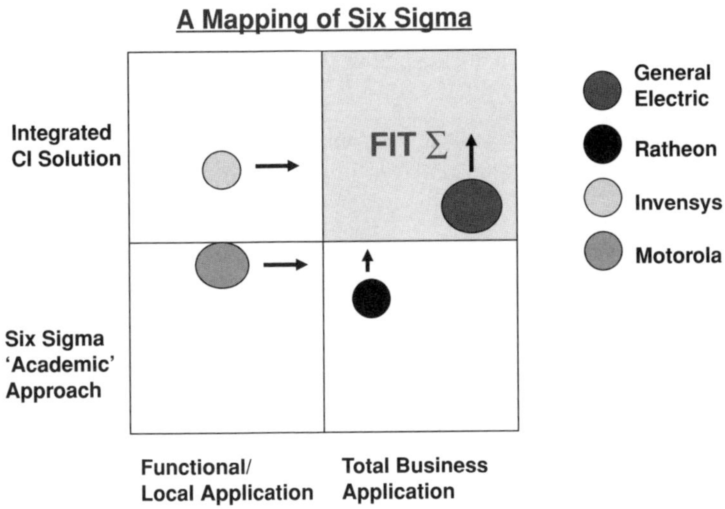

Figure 2.7 A mapping of Six Sigma

In order to ensure that FIT SIGMA spans across all related functions, it is recommended that a value stream mapping is carried out for each key product. This will also help to identify bottlenecks and obvious areas of waste.

We believe that the road map to sustainable success for the companies engaged in the 'pure play' Six Sigma programme constitutes great progress towards a company-wide integration of solutions in the FIT Σ process.

Case example: Integration of Lean and supply chain with Six Sigma at Seagate

Seagate Technology is the world's largest manufacturer of disc drives and HDD recording media. With its headquarters at Scotts Valley, California, the company employs 62,000 people and its turnover in 2000 exceeded $7 billion. The business operates in a market environment of short product life cycle and quick ramp to high volume. The data storage market is growing 10–20% per year and the technology content doubles every 12 months. Volume products remain in production for only six to nine months.

Seagate Springtown (which is part of Seagate Recording) started a supply chain project to improve materials management and develop a strategic vendor relationship. The fabrication plan at Springtown introduced the Lean Manufacturing philosophy that recognises WASTE as the primary driver of cycle time and product cost. Very soon a change had taken place at Springtown and Lean Manufacturing was wholly integrated with the supply chain initiative.

The corporate office at Scotts Valley was rolling out a global Six Sigma deployment programme. The Springtown site followed the Six Sigma training programme and implemented a number of tools and techniques including the Process Map, Sampling Plan, Cause and Effect Analysis and Control Plans, which identified a 'hidden factory'. The less visible defects of this 'hidden factory' included:

- Repeated measurements (in and out).
- Repeated chains (post and pre).

- Transits between manufacturing areas.
- Process steps conducted in 'non-standard operating conditions'.
- High rework on a process.

The Six Sigma methodology proved a key enabler for supply chain/Lean Manufacturing and the integrated programme achieved improved process capability and quality as shown by:

- Increased throughput by 31%.
- Significant impact on capital expenditure due to increased efficiency of existing equipment.
- Lower work-in-progress.
- 80% pass rate on qualifications for vacuum tools (previously 40%).

2.4 FITNESS FOR SUSTAINABILITY

Sustainability, in a general sense, is the capacity to maintain a certain process or state indefinitely. In the context of preserving the environment, it relates to uniting 'the needs of the present without compromising the ability of future generations to meet their own needs' (United Nations, 1987). The sustainability of project outcomes, in the context of FIT SIGMA, is not 'environmental'; rather, it is the longer-lasting stability of the deliverables and processes as illustrated in Figure 2.8.

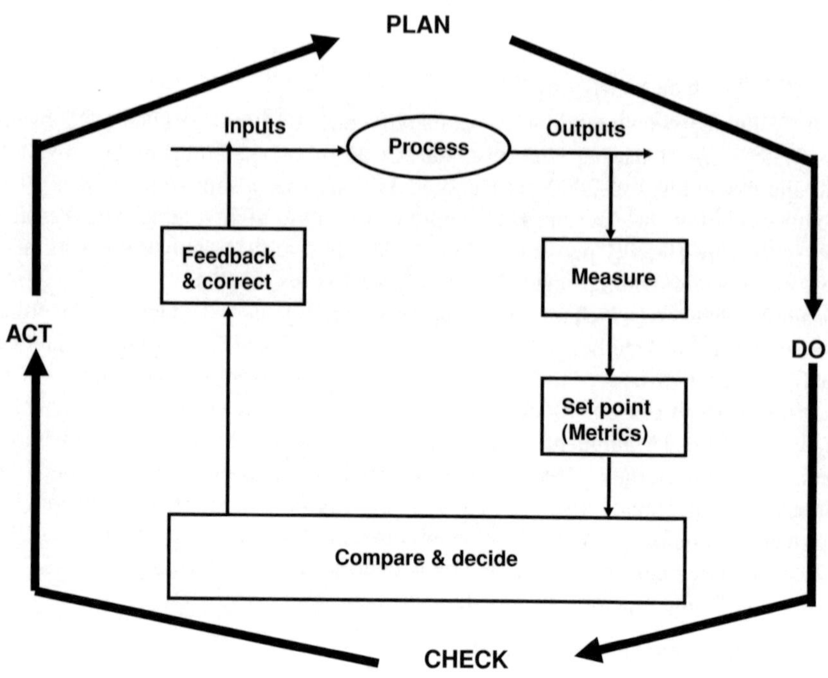

Figure 2.8 A model showing a process towards a longer-lasing outcome in project management (developed from Deming, 1986)

The sustainability of FIT SIGMA project outcomes is comparable to the stability of a process. Stability involves achieving consistent and, ultimately, higher process yields defined by a metric or a set point. This is achieved through the application of an improvement methodology and continuous review. Such stability is ensured by minimising the variation of the set point. The approach of the PDCA cycle (Deming, 1986) enables both temporary and longer-term corrections. The short-term action is aimed at fixing the problem while the permanent corrective steps consist of investigating and eliminating the root causes. This therefore targets the sustainability of the improved process. Thus, a sustainable project is expected to deliver long-lasting outcomes of acceptable quality performance criteria (Mengel, 2008).

The above explanation of sustainable outcome is also in agreement with the following definition of sustained success in British Standards.

> The sustained success of an organisation is the result of its ability to achieve and maintain its objectives in the long-term. The achievement of sustained success for any organisation is a complex and demanding challenge in an ever-changing environment.
>
> BS EN ISO 9004:2009

Having steered your way through the challenges of Six Sigma and Lean Sigma programmes over recent years, you are quite possibly very proud of its achievements in terms of results and the leanness of operations. You certainly deserve to congratulate yourself on attaining your goal, but it must be said that this is just the beginning. Realistically, you have only just embarked upon your path of success.

In a Lean programme, the reduction of overheads and elimination of non-value-added activities are all excellent accomplishments in themselves. However, no Lean practitioner can afford to be complacent. In layman's terms they can be likened to someone who is keen to lose weight and so 'crash diets' without incorporating an appropriate fitness programme to back up their new regime. Sure, such a dramatic reduction in calorific intake will produce speedy results, but this will be in the short term only. The target weight, once attained, will be very difficult to uphold without making more far-reaching, holistic changes to nutrition, exercise and lifestyle. Experts always tell us that the only way to achieve and maintain a goal weight long term is to completely review every aspect of our lives. We all know that crash diets have an immediate but unsustainable effect – in the end, we are sure to break it, however noble our intentions!

This analogy is one we can all understand and bear in mind when considering how to achieve long-term success and maintain goals that are important to us. Thus, in a FIT $\sum$ programme the sustainability of performance is instilled right through the process and not just after the implementation of the deployment plan.

The fitness for the sustainability of performance is underpinned by four key processes:

1. Performance management.
2. Senior management review.
3. Self-assessment and certification.
4. Knowledge management.

2.4.1 Performance Management

The point that the success of Six Sigma is highly focused on measurements, both statistical and savings, makes performance management a logical and essential component of the programme.

In the context of FIT Σ, we address some relevant issues including what we measure, when we measure and how we measure.

There is little doubt, even in the present environment of advanced information technology, that a company's performance is governed by quarterly or annual financial reports. These accounts create an immediate impact on the share value of the company. As the financial reports are linked, an accounting model developed a long time ago for site-centric activities. The majority of performance measures are still rooted to the traditional accounting practice. The senior managers of a company are usually driven to improve the share values which affect their personal share options in the firm. Thus, we find the traditional accounting model is still being used, even by information-age companies.

In addition to reporting basic financial measures (e.g. Sales Value, Net Profit, Equity, Working Capital and Return on Investment), other conventional measures have been extended to assess Customer Service (market effectives) and Resource Utilisation (operations efficiency). Wild (2002) argues that the three aspects of customer service – specifications, cost and timing – can be measured against set points or targets. Given many resources as input to a process, resource utilisation can be measured as 'the ratio of useful output to input'. Resource utilisation is cost-driven while the objective of customer service is 'value added' to the business.

The models of financial accounting, customer service and resource utilisation may also be applicable to some areas of FIT Σ, but these may not incorporate the key aspect of the programme.

Kaplan and Norton (2004) argued that 'a valuation of intangible assets and company capabilities would be especially helpful since, for information age companies, these assets are more critical to success than traditional physical and tangible assets'. They have created a new model called 'The Balanced Scorecard', as illustrated in Figure 2.9.

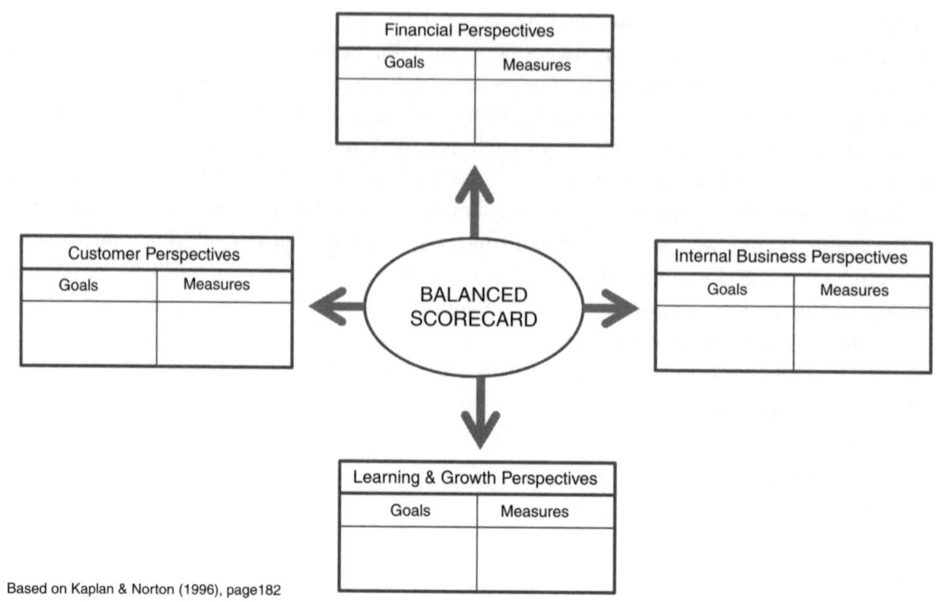

Figure 2.9 Kaplan and Norton's Balanced Scorecard

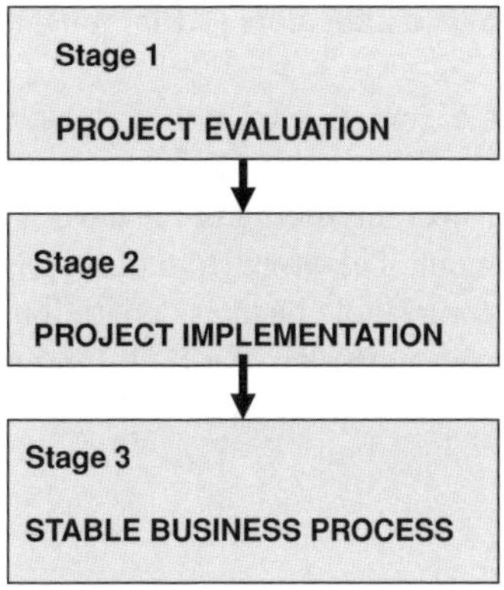

Figure 2.10 The stages of performance management

The Balanced Scorecard retains traditional financial measures, customer services and resource utilisation (internal business process) and includes additional measures for learning (people) and growth (innovation). This approach complements measures of past performance with drivers for future development. The Balanced Scorecard can be applied to a stable business process following good progress with the FIT Σ programme.

Performance management in FIT Σ should also be 'fit for the purpose' and the appropriate metrics should depend on the stages of the programme. There are three key stages of a FIT Σ initiative in the context of measuring its performance, as shown in Figure 2.10.

As discussed earlier, larger projects in a FIT Σ programme are selected based upon an organisation's strategic goals and requirements. The viability of the project is then established dependent upon certain quantifiable criteria including ROI (return on investment). At the project evaluation stage of a FIT Σ initiative, similar criteria should prevail. Although attempts must be made to show an 'order of magnitude' of ROI data, the emphasis should be focused more on strategic goals and requirements.

The measurement process at the project implementation stage is basically the monitoring of the key factors considered during the project evaluation phase. The following six elements are suggested.

The first factor is to determine the project's value to the business, which can be reflected in the company's overall financial performance. This aspect can be applied by monitoring the savings on a monthly basis.

The second characteristic is the resources required. If resources are outsourced then this cost is measured and monitored. The timescale of the project is also included in this factor.

The third element is the metrics that may be required to monitor the performance of specific large projects. Examples of this factor are DPMO (defects per million opportunities) and RTY (rolled throughput yield).

Fourth, it is necessary to monitor the impact of the project on the external market: whether there is an eroding customer service or sales revenue as a result of key company resources deployed in the project.

The fifth phase is to ensure that the FIT Σ initiative continues to align with the overall mission and strategy of the business.

Finally, selective key performance indicators (KPIs) must be established for the next stage of the stable business process.

The transition period of a FIT Σ initiative from the project juncture to a stable business operation is often difficult to pinpoint. The main reasons for contribution to this blurred situation are due to the relatively short project duration and the need for continuous modifications in a dynamic technological environment. Therefore, five of the six factors (excluding the second dynamic) should be monitored for a stable business operation. However, additional emphasis should be given by focusing on the sixth step, the KPIs, and gradually all factors can then be incorporated in selective KPIs. A customised Balanced Scorecard should be appropriate for a stable business process.

Case example: Monitoring performance during a Six Sigma programme at Dupont Teijin Films

Dupont Teijin Films is a global polyester films business with manufacturing sites in the USA, Europe and Asia. The company was created following the acquisition of Teijin Films of Japan by Dupont. DTF is a market leader but experiencing tough competition from new entrants.

As part of the corporate Six Sigma programme, the Wilton Site of DTF in Middlesborough, UK, started the deployment plan in 1999. The main objectives of the programme included:

- Increased capacity.
- Improved material efficiencies.
- Cost reduction.
- Increased revenue by higher sales volume.

The site project team followed a methodology of 'successful implementation' in three key categories – 'Doing the Right Work' (Process); 'Doing the Work Right' (Efficiency); and 'Creating the Right Environment' (Education and Culture).

Within the category of 'Doing the Work Right' the team introduced:

- Input Metrics.
- Output Metrics.
- Tracking Profile.

The Input Metrics included the number of Black Belts trained and the total number of people trained. The Output Metrics covered:

- $ saved;
- number of projects per annum;
- quality index;
- CTQ flowdown;
- COPQ;
- strategic lining.

The project team followed an internal self-assessment process every quarter based on a 'Do Right Work Checklist'. This comprised 24 questions covering Customer Alignment, Business Alignment, Process Baselining and Project Selection.

2.4.2 Senior Management Review

A recurring challenge exists for companies who have invested significant time and resources in implementing proven improvement plans such as Six Sigma. Put simply, this is how to ensure their sustainable performance beyond the duration of a one-off corporate exercise. The annual review of the change programme during the budget planning is ineffective because 12 months is a long time in a competitive marketplace. We can use the analogy of travelling in a car to understand this issue. In order to steer the benefits of the programme and the business objectives to a sustainable future, the senior managers who are in the driving seats must have a clear view of both the front screen and the rear view mirrors. In addition, they must look at them as frequently as possible to decide on their direction and optimum speed.

In recent years the pace of change in technology and marketplace dynamics has been so rapid that the traditional methodology of monitoring actual performance against predetermined budgets set at the beginning of the year may no longer be valid. It is fundamental that businesses are managed based on current conditions and up-to-date assumptions; there is also a vital need to establish an effective communication link, both horizontally across functional divisions and vertically across the management hierarchy, to share common data and decision processes. One such solution to these continuous review requirements is sales and operations planning (S&OP).

Sales and operations planning has become an established company-wide business planning process in the Oliver Wight MRPII methodology (Ling *et al.*, 1988). The diagram in Figure 2.11 shows the five steps in the process that will usually be present. The course of action can be adapted to specific organisational requirements.

New Product Review (Step 1). Many companies follow parallel projects related to the new products in R&D, Marketing and Operations. The purpose of this review process in Step 1 is to examine the different objectives of various departments at the beginning of the month and resolve new product-related assumptions and concerns. The issues raised will impact upon the demand plan and the supply chain at a later stage of the process.

Demand Review (Step 2). Demand planning is more of a consensus art than a forecasting science. Demand may change from month to month depending on market intelligence,

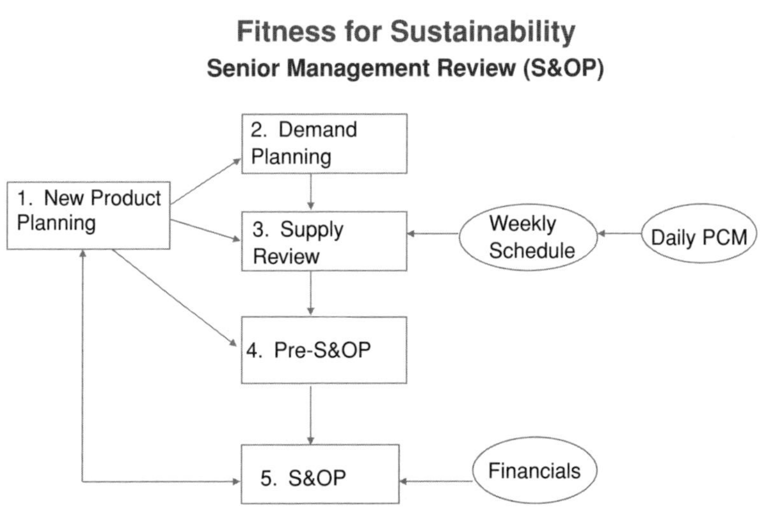

Figure 2.11 Senior management review process (S&OP)

customer confidence, exchange rates, promotions, product availability and many other internal and external factors. This review at the end of the first week of the month, between Marketing, Sales, IT and Logistics, establishes agreement and accountability for the latest demand plan identifying changes and issues arising.

Supply Review (Step 3). In the current climate of increasing outsourcing and supply partnership, the capacity of supply is highly variable and there is a need to ensure the availability and optimisation of supply every month. This review, usually on the second week of the month, between Logistics, Purchasing and Production, establishes the production and procurement plans and raises capacity, inventory and scheduling issues.

Reconciliation Review (Step 4). Problems would have been identified in previous reviews of new products, demand and supply. The reconciliation step goes beyond the balancing of numbers to assess the business advantage and risk for each area of conflict. This review looks at issues from the business point of view rather than departmental objectives. This is also known as the pre-S&OP review, and its aim is to minimise issues for the final S&OP stage.

Senior Management Review (Step 5). Senior managers or board members, with an MD or CEO in Chair, will approve the plan that will provide clear visibility for a single set of members driving the total business forward. The agenda includes the review of key performance indicators, business trends of operational and financial performance, issues arising from previous reviews and corporate initiatives. This is a powerful forum to adjust business direction and priorities. This is also known as the sales and operations planning (S&OP) review.

At each process step the reviews must address a planning horizon of 18–36 months in order to make a decision for both operational and strategic objectives. There may be a perceived view that S&OP is a process of aggregate/volume planning for supply chain. However, it is also a top-level forum to provide a link between business plan and strategy. The continuous improvement and sustainability of company performance by a FIT Σ programme can only be ensured in the longer term by a well-structured S&OP or Senior Management Review Process. The results and issues related to FIT Σ should be a regular item in the S&OP agenda. Figure 2.12 illustrates how a hierarchy of KPIs can be applied and cascaded across the review process.

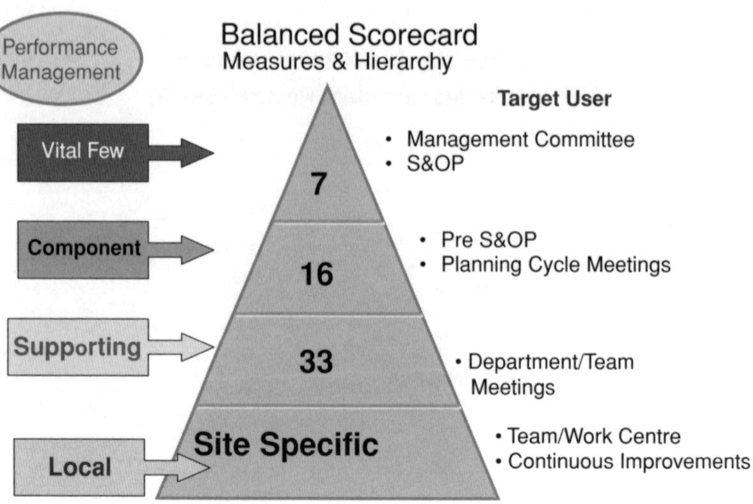

Figure 2.12 Balanced Scorecard: measures and hierarchy

Case example: S&OP ensures sustainable performance at GSK Turkey

GlaxoSmithKline Turkey (GSK Turkey, previously known as GlaxoWellcome Turkey) was awarded MRPII 'Class A' certification in 1999 by business education consultants Oliver Wight Europe.

GSK Turkey launched a programme (known as EKIP) in January 1998 to improve company-wide communications and sustain a robust business planning process using MRPII 'best practice' principles.

Beginning in September 1998, the company improved and sustained a customer service level running at 97% and inventory turnover of around 5.0. The sales turnover in 1998 increased by 20% in real terms in spite of some supply shortfall from the corporate network in the first half. This was an even greater achievement in view of the adverse economic and political conditions of Turkey at the time. GSK Turkey was recognised as a major business in the pharmaceutical giant GSK Group and the business plan for 1999 was aiming at a turnover of US$110 million.

As part of the MRP II Class A programme, GSK Turkey installed an S&OP process which was underpinned by a set of business planning meetings at various levels. In spite of the GW and SB merger and the corporate Lean Sigma initiative, the S&OP process has since been continued by the company every month, on an ongoing basis.

The vigour of the S&OP process, which is championed by the Managing Director, has helped the company to sustain and improve both the business benefits and its communication culture. This was particularly useful when they were challenged by a number of initiatives in hand, including:

- Transfer of office.
- Rationalisation of factory and warehouse.
- Corporate Lean Sigma programme.
- Merger of GlaxoWellcome and Smith Kline Beecham.

2.4.3 Self-assessment and Certification

In order to maintain a wave of interest in the quality programme and also to market the competitive advantage that quality undoubtedly affords, many companies channelled their efforts in two respects. They directed their energies to the pursuit of an approved accreditation such as ISO 9000, or to a prize such as the Malcolm Baldridge Award (in the USA) or derivatives of the Baldridge Award (in other countries). The process of certification and awards has had a chequered history. After a peak during the early 1990s, the Baldridge Award gradually lost its impact in the USA and companies such as GE or Johnson and Johnson started developing their own customised quality assessment process. Encouraged by customer demand for the ISO stamp of approval, there was a rush for ISO 9000 certification in the 1990s. However, many firms became disillusioned by the auditors' focus on ensuring compliance with mainly current procedures without necessarily improving standards. A number of consultancy companies attempted to introduce their own awards to progress an improvement programme (e.g. 'Class A' by Oliver Wight).

It is essential to incorporate a self-assessment process in a FIT Σ programme in order to sustain a performance and improvement culture. Put simply, there are two choices: either select

Table 2.3 Pros and cons of the self-assessment options

Option	Pros	Cons
Standard accreditation	Proven process Known to customers and suppliers Trained auditors and consultants available External networking	Too generic to fit business Invasion of auditors and consultants More expensive Not improvement driven
Customised self-assessment	Process ownership Customised to business needs Improvement orientated Common company culture In-house knowledge based Enables self-assessment	Lack of external benchmark Time to develop and pilot

an external certification or develop your own checklist based on proven processes. Table 2.3 highlights the relative pros and cons of these two options.

There are several examples where a company achieved an external award based on a set of criteria but without improving business performance. There are also cases where, after a burst of initial publicity, the performance level and pursuit for excellence were not maintained. If the process is not underpinned by self-assessment then the award will gradually lose its shine, just like lack of maintenance of an expensive new car. We therefore recommend that a FIT Σ programme should adopt a self-assessment process developed from proven procedures. Two such courses of action are described below: 'EFQM' and 'Total Solutions'.

2.4.3.1 EFQM (European Foundation of Quality Management)

The EFQM accolade is derived from America's Malcolm Baldridge National Quality Award. There are similar tributes available in other countries, such as the Canadian Excellence Awards and the Australian Quality Award.

The EFQM Award was established in 1991. It is supported by the European Union and countries within the EU have their own support unit (e.g. British Quality Foundation in the UK). As shown in Figure 2.13, the ERQM model provides a set of checklist questionnaires under nine categories, each providing a maximum number of points. They are:

1	Leadership	100 points
2	People Management	90 points
3	Policy and Strategy	80 points
4	Resources	90 points
5	Processes	140 points
6	People Satisfaction	90 points
7	Customer Satisfaction	200 points
8	Impact on Society	60 points
9	Business Results	150 points
TOTAL		**1000 points**

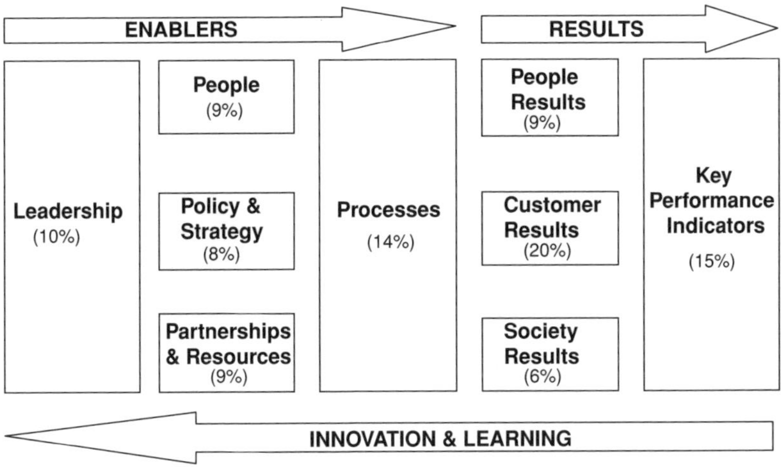

Figure 2.13 The EFQM model

The first five categories (Leadership to Process) can be seen as 'enablers', while the remaining four categories are 'performance'-related.

2.4.3.2 Total Solutions

This holistic approach of self-analysis covering all aspects of the business has been described in detail in *Total Manufacturing Solutions* (Basu and Wright, 1998). As shown in Figure 2.14, Total Solutions enables self-assessment against 20 defined areas ('foundation stones') to identify areas of improvement for achieving the full potential of the business.

The business is built from the foundation stones up, and consists of the 'six pillars' of total solutions. There are 200 questions in the checklist, 10 questions for each foundation stone. These pillars are comprised as follows:

- Marketing and Innovation.
- Supply Chain Management.
- Environment and Safety.
- Facilities.
- Procedures.
- People.

Although the checklist is aimed at manufacturing operations, it can easily be adapted to service operations. A 'spider diagram' can be constructed from scores of each foundation stone to highlight the current performance profile and gaps (see Figure 2.15).

Our recommended methodology of the self-assessment in a FIT Σ programme comprises the following features.

1. Establish the policy of external certification or customised self-assessment in line with the company culture and business characteristics.
2. Develop or confirm the checklist of assessment.

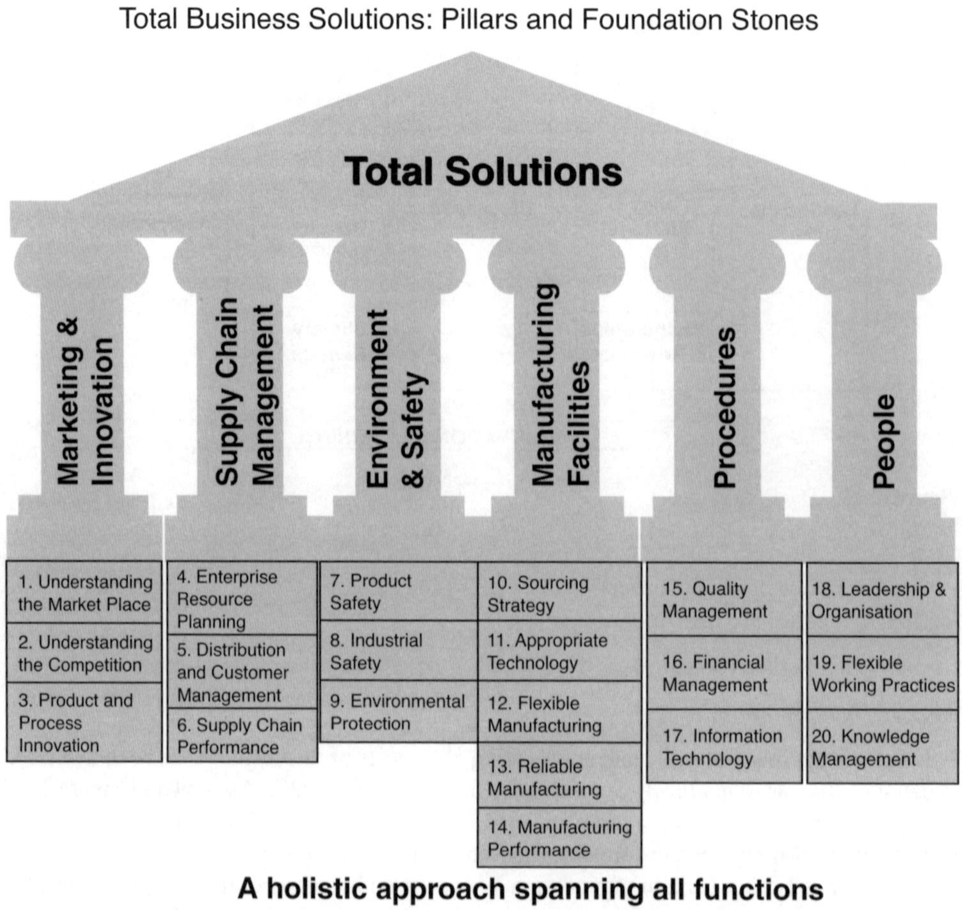

Figure 2.14 Total Solutions: a holistic approach

3. Train internal assessors in the common company assessment process (one assessor for every 500 employees as a rough guide). These assessors should also carry out normal line or functional duties.
4. Train experts (Black Belts) and department managers in the self-assessment checklist and process.
5. Carry out quarterly self-assessment by departmental managers.
6. Ensure six-monthly (at the initial stage) and annual (at the later phase) assessments by internal assessors.
7. Analyse gaps and implement measures to minimise these shortcomings.
8. Consider corporate awards, depending on the performance attained, by the CEO.
9. Review the checklist with the change of business every two years.
10. Consider external accreditation if it adds value to the business.

The above methodology is applicable to all types of business, both manufacturing and service, and all sizes of operations whether large, medium or small. A larger organisation is

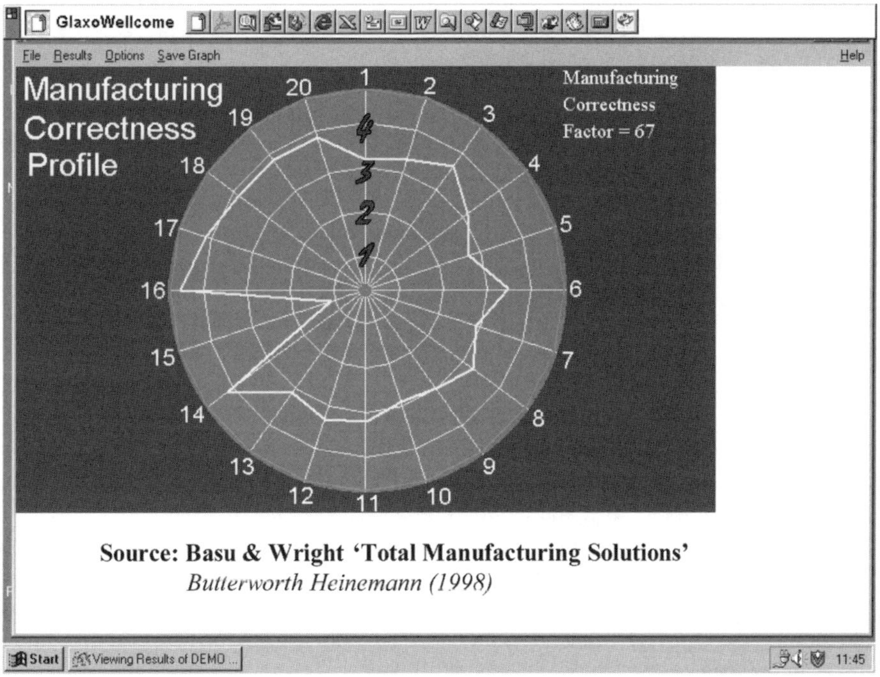

Figure 2.15 Manufacturing correctness profile

likely to possess its own resources to develop and maintain the process. A smaller organisation may require the assistance of external consultants to develop the procedure.

Case example: Janssen-Cilag applies 'Signature of Quality' for continuous self-improvement

Janssen-Cilag is the pharmaceutical arm of the Johnson and Johnson Group with their European Head Office based in High Wycombe, Buckinghamshire. The origins of the company lie as far back as the 1940s, with three firms initially in existence: Ortho Pharmaceutical in the UK, Cilag Chemie in Switzerland and Janssen Pharmaceutica in Belgium. The merger was completed in 1995 and Janssen-Cilag is now among the top 10 pharmaceutical companies in the world. The company markets prescription medicines for a range of therapeutic areas of gastroenterology, fungal infections, women's health, mental health and neurology.

The commitment of the company to the values and standards laid out in 'Our Credo' drives management to strive continually for excellence in a number of overlapping areas. Based upon the principles of the Baldridge Award, the Quality Management team of Janssen-Cilag developed a self-assessment process known as 'Signature of Quality' (SoQ). The process is supported by a checklist on a carefully constructed questionnaire in five interdependent areas:

- Customer Focus
- Innovation

- Personnel and Organisational Leadership
- Exploitation of Enabling Technology
- Environment and Safety

SoQ is managed as a global process from the US office and each site is encouraged to prepare and submit a comprehensive quality report meeting the requirements. The assessment is carried out by specially trained Quality Auditors and a site may receive a SoQ Award based upon the results of the assessment.

SoQ has been reported to be successful in Janssen-Cilag as a tool for performing a regular 'health check' and as a foundation for improvement from internal benchmarking.

2.4.4 Knowledge Management

Almost 400 years ago, Francis Bacon stated that 'Knowledge is power'. Peter Drucker wrote in the *Atlanta monthly*, 11 September 1995, 'Knowledge has become the key economic resource and the dominant, if not the only, source of comparative advantage'.

Webster's dictionary defines knowledge as 'familiarity or understanding gained through experience or study'. In the context of a FIT Σ initiative, an essential sustainable driver of performance management is the sharing of knowledge and best practice. Although not explicit in his 14-point philosophy, Deming advocated the principle 'to find the best practices and train the worker in that best way'.

The key tenets of knowledge management in FIT Σ methodology are:

1. Systematically capture knowledge from proven 'good practices'.
2. Select examples of 'best practices' based upon added value to the business.
3. Do not differentiate between the sources, regardless of level of technology or economic power.
4. Inculcate knowledge-sharing between all units.

The essential ingredient for benefiting from knowledge management is the establishment of a 'learning organisation' culture. The key element here is the faith of the participants in the process. Unless members at all levels of a company involved in sharing knowledge believe that their business can benefit from it, then the exercise has little value. If a company thinks that they already know the best way or that the 'best practice' is actually not appropriate to their circumstances, then sustainable improvement just will not happen. The development of a 'learning organisation' culture does not, of course, happen overnight – it takes time, it requires the appropriate infrastructure to be in place. Our experience is that time and money spent in knowledge management is also invested in the most valuable resource of competitive advantage – people. The support structure for such a knowledge-sharing process should include:

- A champion to act as a focal point to co-ordinate the process.
- A regular best practice forum to learn from each other and to allow networking.
- Internal and external benchmarking to assess targets and gaps.
- Continuous communication through websites, newsletters, videos and 'visual factors'.

Case example: GE Capital shares best practice of 42 branches

General Electric Inc., with its global business of over 120 billion per annum, has been voted by *Fortune* as 'the most respected company'. GE is also known as the 'Cathedral of Six Sigma' and the high profile of the programme under the leadership of Jack Welch has been well publicised. GE licensed Six Sigma technology in 1994 from the Six Sigma Academy and rolled out the programme worldwide, achieving $2 billion savings in 1999.

GE Capital is the financial services arm of GE and accounts for approximately 40% of the group turnover. The success of Six Sigma in GE Capital has been a testament to the progress in service operations. One good early success story of GE Capital relates simply to the sharing of good practice.

GE Capital fielded about 300,000 calls a year from mortgage customers who had to use voice mail or call back 24% of the time because the employees were busy or unavailable. A Six Sigma team found that in actual fact, one of their 42 branches had a near-perfect percentage of answered calls.

'The team analysed its systems, process flows, equipment, physical layout, and staffing, and then cloned it to the other 41. Customers who once found us inaccessible nearly one quarter of the time now had a 99.9 per cent chance of getting a GE person on the first try.' (Jack Welch, 2001)

2.5 SUMMARY

FIT SIGMA is a natural extension of the third wave of the Quality Movement offering a historically proven process to improve and sustain the performance of all businesses, both manufacturing and services, whether large or small.

FIT SIGMA is not a statistic. It is both a management philosophy and an improvement process. The underlying belief is that of a total business-focused approach underpinned by continuous reviews and a knowledge-based culture to sustain a high level of performance. In order to implement the FIT SIGMA philosophy, a systematic approach is recommended. The process is not a set of new or unknown tools; in fact, these tools and cultures have been proven to yield excellent results in earlier waves.

The differentiation of FIT SIGMA is the process of combining and retaining success factors. Its strength is that the process is not a rigid programme in search of problems, but an adaptable solution for a specific company or business.

Small wonder then that FIT SIGMA can be seen to offer new and exciting possibilities in the field of operational excellence. There is no magic formula in a new name or brand; what counts are the underlying total business philosophy process and culture of FIT SIGMA.

In the following chapters, the applications of FIT SIGMA in organisations of different sizes and sectors will be described. It is to be noted that not all applications have adopted the formal name of FIT SIGMA in their programmes, but they illustrate the flexibility and tenets of FIT SIGMA in each situation. For example, for a large manufacturing organisation the approach of FIT SIGMA is similar to that of Lean Sigma with the added principle of 'fitness for sustainability'.

3

DMAIC Methodology for FIT SIGMA

3.1 INTRODUCTION

DMAIC (Define, Measure, Analyse, Improve, Control) is a structured problem-solving methodology used extensively in Six Sigma projects. The DMAIC methodology is to be used when a process is in existence at your company but is not meeting customer specification or is otherwise not performing adequately. It combines the systematic data-driven approach of classical industrial engineering with the rigour of the life cycle in project management. The effectiveness of this approach cannot be doubted. However, a full application of DMAIC is time-consuming and can be too costly for some applications in smaller projects. Therefore, it becomes necessary to question its proportionality of application. This chapter aims to address this enquiry by using the FIT SIGMA theme of 'fitness for the purpose'.

There are three approaches of the DMAIC methodology for FIT SIGMA in order to satisfy the theme of 'fitness for the purpose'.

1. DMAIC Full.
2. DMAIC Lite.
3. KAIZEN Event.

DMAIC Full is applicable to:

- Complex multifunctional projects.
- Potential savings over £250,000.
- Project duration likely to be over three months.
- Suitable for large organisations with adequate training and programme budget.
- A Black Belt led project.

DMAIC Light is applicable to:

- Departmental projects.
- Project duration likely to be less than three months.
- Suitable for SMEs.
- Confidence limit less than 95%.
- A Green Belt led project.

KAIZEN Event is applicable to:

- A predominantly Lean project when obvious sources have been identified.
- When implementation risk is minimal.
- When results are needed as soon as possible.
- Suitable for both large organisations and SMEs.

3.2 DMAIC FULL

3.2.1 Background

During the early part of the 1980s, total quality management (TQM), especially in the USA and Europe, concentrated on the application of quality management principles and SPC tools within all aspects of the organisation. The focus was upon their integration with key business processes. There were considerable successes in communication and culture but despite this, something was missing – the rigour of project management principles. DMAIC was introduced by Motorola as the life cycle discipline for Six Sigma projects in the late 1980s. Since then, DMAIC has become the essential component of all Six Sigma initiatives and training programmes. In order to emphasise the sustainability of gains during the Control phase, some companies (e.g. Dow Chemical) have added the process of Leverage (L) after Control and thus extended DMAIC to become DMAICL.

While Deming's PDCA (Plan, Do, Check, Act) cycle has been used extensively in the development and deployment of quality policies, DMAIC has added the rigour of project life cycle to the implementation and close-out of Six Sigma projects. Figure 3.1 shows the relationship between DMAIC with PDCA and a typical project life cycle.

3.2.2 Definition

DMAIC refers to a data-driven life cycle approach to Six Sigma projects for improving processes and is an essential part of a company's Six Sigma programme. DMAIC is an acronym for five interconnected phases: Define, Measure, Analyse, Improve and Control. The simplified definitions of each phase are:

Define by identifying, prioritising and selecting the right project.
Measure key process characteristics, the scope of parameters and their performances.
Analyse by identifying key causes and process determinants.
Improve by changing the process and optimising performance.
Control by sustaining the gain.

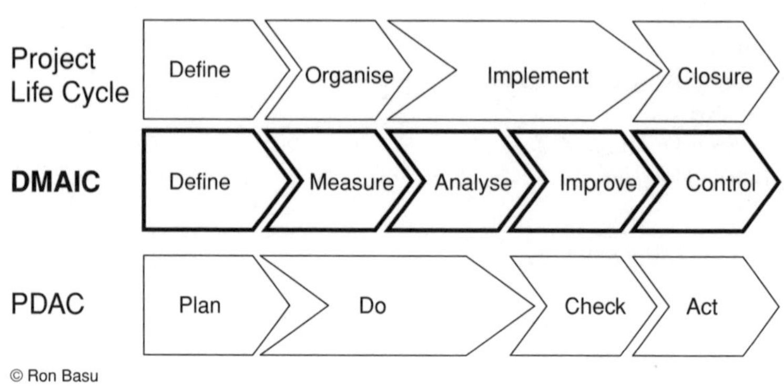

Figure 3.1 DMAIC life cycle

3.2.3 Application

The tools of Six Sigma and operational excellence, as described in this book, are applied most often within the framework of DMAIC. As such, DMAIC is an integral part of a Six Sigma initiative.

DMAIC is also used to create a 'gated process' for project control. For example, the criteria for a particular phase are defined and the project is reviewed; if the criteria are met then the next phase starts (see Figure 3.2).

As a summary of the application of the DMAIC technique, if you cannot define your process you cannot measure it. That means if you cannot express the data you are not able to utilise DMAIC in your development actions. Therefore you cannot improve and sustain the quality.

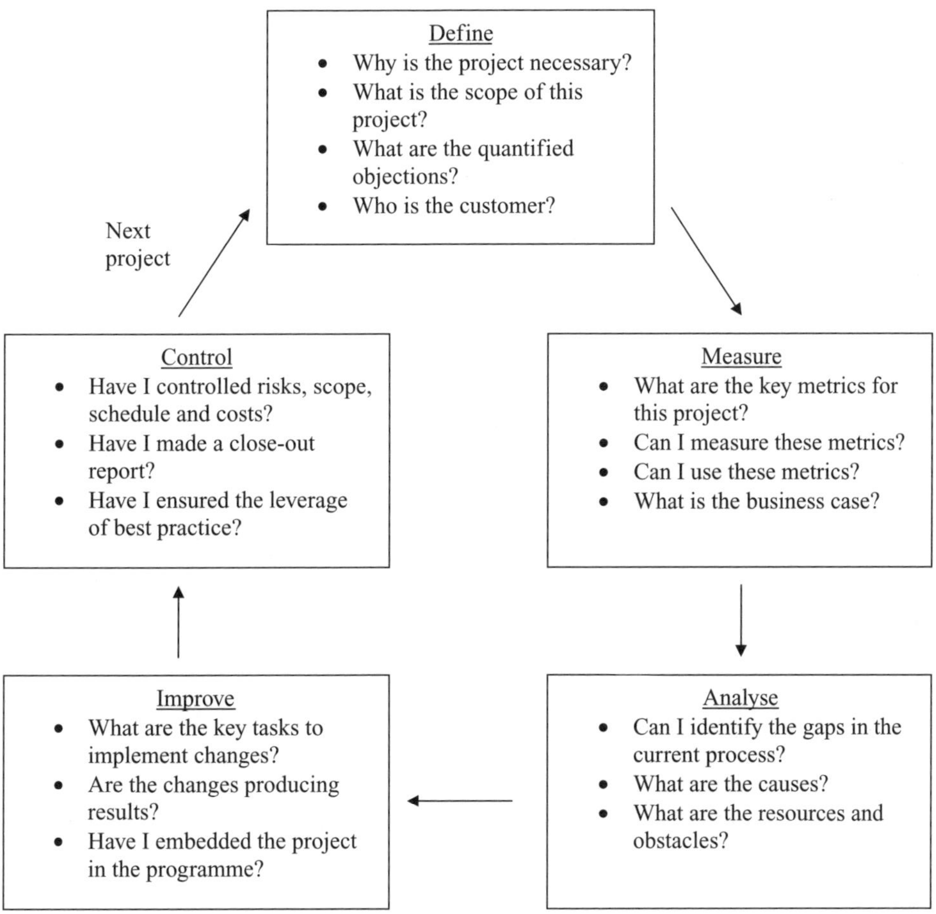

Source: Adapted from www.pyzdek.com.

Figure 3.2 The DMAIC cycle as a 'gated process'

3.2.4 Basic Steps

The key steps of the DMAIC process are well-defined. Each phase is required to ensure the best possible results. The process steps are:

1. Define. In this initial phase, it is important to characterise the purpose, scope, objectives and expectations, resources and time line of the project. The customers, their CTQ issues and the core business process involved are clearly identified at this stage.

The tools used for 'Define' include benchmarking, project charter, process mapping and SIPOC.

The deliverables of this phase are:

- Project definition report (project charter).
- Customers and CTQs defined.
- Business process mapped.
- Fully trained team formed.

2. Measure. At the start of any journey, the most important thing to know is where you are going. 'Define' provides this road map. However, even if you have a reliable route you will get lost unless you know where you are now. The 'Measure' period pinpoints your current location by defining the baseline. One or more CTQs are measured to elicit sufficient data from the process under investigation. At a minimum, consider the mean performance and some estimate of variation.

The tools at this stage are process flow charts, selected SPC tools, Pareto analysis and run charts.

The main deliverables of this phase include:

- A proven set of metrics to measure the project progress and objectives.
- Process capability and Sigma baseline.

3. Analyse. Once the baseline performance is documented, the objective of the Analyse stage is to validate the root causes of problems. You should be able to compare the various options to determine the most attractive alternative. Striking the appropriate balance between details of the analysis is a key success factor.

Any number of tools and tests can be used. The useful tools for Analysis are the cause and effect diagram, SPC tools and the 'five whys'. For an advanced application, the FMEA and DOE techniques are used extensively.

The important deliverables of this phase are:

- Special and common causes of variance.
- Prioritise opportunities to improve.

4. Improve. During the 'Improve' step, ideas and solutions are put to work. There must be checks to ensure that the desired results are achieved and some trials and experiments may be required in order to find the best solution.

The tools and techniques at this stage include mistake-proofing, DOE and simulation.

The main deliverables of this phase are:

- The optimum process for improvement.
- The key tasks required (work breakdown) to implement the improvement.

5. Control. At the final 'Control' phase the stability and reliability of the process is assured and the final capability is determined. The documentation for leveraging the best practice is also part of the Control juncture.

The tools and techniques used at this step include SPC, FMEA and benefits tracking. The main deliverables are:

- Cost–benefit analysis.
- Close-out report.

Summary of steps in DMAIC

Define phase
- Management initiatives
- Problem definition
- Collect voice of customer (VOC)
- Develop big picture map (SIPOC)
- Sign off project charter and launch team

Measure phase
- Define performance standards
- Value stream mapping
- Measurement system analysis
- Monitoring the process
- Establish process capability
- Gate review and update project charter

Analyse phase
- Determine CTQ characteristics
- Pareto analysis
- Cause and effect diagram
- Brainstorming
- Identify variation sources
- Gate review and update project charter

Improve phase
- Generate potential solutions
- Develop and implement pilot solution
- Establishing five S system
- Confirm attainment of project goals
- Gate review and update project charter

Control phase
- Institute ongoing metrics and control charts
- Share the lessons learnt
- Document standard operating procedure
- Mistake proofing exercise
- Finalise project and hand over to process owner
- Audit the results
- Institute sustainability plan

3.2.5 Worked-out Example

Although DMAIC is a data-driven and quantitative technique, it is also a methodology that may not produce unique output from a given set of input data. We have therefore chosen an actual case (from Dow Chemical) to illustrate a worked-out example.

3.2.5.1 Define

The Dow Chemical Company began using Six Sigma in 1998 to improve the operations of its subsidiary – the Film Tec Corp. of Minneapolis – a manufacturer of membranes that was having difficulty in meeting customer demand.

Membrane quality is determined by two criteria: (a) flux, or the amount of water the membrane lets through during a given period and (b) how much impurity is removed from the water.

3.2.5.2 Measure

Membrane elements were tested prior to shipping to ensure the prescribed quality standards. The specifications were assured but the speed with which his customers were serviced suffered. The shortfall of customer service was costing Film Tec approximately $500,000 a year.

3.2.5.3 Analyse

Six Sigma tools and methodologies were used at Film Tec to reduce product and process variation. Statistical analysis was used to identify variables that affected membrane flux most significantly.

One of the variables identified for improvement was the concentration of a chemical component used in the manufacturing process. The problem stemmed from the inconsistencies in concentration caused by the interruptions in feeding the chemical in the manufacturing process. The feeding was done from a movable container that was replaced every day. When the empty container was replaced by a full one, the chemical did not always reach the process area. Empty containers were often not noticed.

3.2.5.4 Improve

To reduce the variation, an inexpensive reservoir was added to fill the chemical while containers were exchanged. Additionally, a level transmitter with an alarm was installed to alert operators to containers that were nearly empty.

The improvements have been significant. For one of Film Tec's water membranes, the standard deviation of the product out of spec. was reduced from 14.5% to only 2.2%. This resulted in several benefits to Film Tec and its customers, including the bonus that membranes were made available to customers faster than before.

3.2.5.5 Control

To sustain the gains from the project, the Six Sigma team made additional changes. Prior to DMAIC, numbers used to track trends were displayed in tabular form. However, the tables were difficult to read and had little impact on monitoring. Now measurements are displayed on Excel charts that illustrate graphically and with visual immediacy the trends in flux on the finished membrane. The results and learnings from this project were incorporated into a report and posted on the knowledge management website of Dow Chemical.

3.2.6 Benefits

The success of DMAIC is so embedded in the success of Six Sigma initiatives that DMAIC is often regarded as being synonymous with Six Sigma. The training programmes (e.g. Black Belt and Green Belt) are also centred around DMAIC. The grouping of the tools in this book is also roughly structured along the lines of DMAIC. The main benefits of DMAIC include:

1. It provides a systematic approach which is common to everyone involved in the Six Sigma programme.
2. DMAIC represents the life cycle of each Six Sigma project – thus imparting a discipline and rigour of project management to achieve its objectives.
3. The results-orientated approach of DMAIC ensures the tracking of benefits and savings.
4. Unlike other advanced quantitative techniques of Six Sigma (viz. DOE, QFD), DMAIC is easy to follow for all members of the project team.

3.2.7 Pitfalls

DMAIC is not immune to pitfalls, however. These include:

1. The critics of Six Sigma often argue that there is nothing new in DMAIC. Similar results, they point out, could be achieved by Deming's PDAC cycle or the basic steps of classical Industrial Engineering.

Industrial Engineering	Six Sigma
Select Record	Define
Measure	Measure
Analyse	Analyse
Develop Implement	Improve
Maintain	Control

2. It is conceivable that some less complex projects (or the so-called 'low-hanging fruits') may be subjected to the rigour of DMAIC more than is necessary, and thus delay the realisation of results.

3.2.8 Training requirements

The training process of DMAIC can be carried out at three levels:

1. Awareness of DMAIC principles in a one-day workshop.
2. Detailed training of DMAIC (for Green Belts) in one week.
3. Training and application of DMAIC as part of Black Belt training, over five or six weeks. This would include hands-on training of DOE, FMEA and QFD.

DMAIC is an integral part of Six Sigma. It is both systematic and fact-based and provides a rigorous framework of results-orientated project management. The methodology may appear to be linear and explicitly defined, but it should be noted that the best results from DMAIC are achieved when the process is flexible, thus eliminating unproductive steps. An iterative approach may be necessary as well, especially when the team members are new to the tools and techniques.

3.3 DMAIC LITE

3.3.1 Background

DMAIC is a powerful data-driven process within the Six Sigma programme. The framework works very well in a wide variety of situations when applied after the adequate training of the process and the associated tools. Therefore, using DMAIC involves time and expense and also the leadership of a Black Belt. In a complex problem the causes and solutions are not obvious. When the solution affects many departments or functions, the risks of implementation are likely to be high. Under these conditions, when the problem is complicated or when the risks in the solution are great, the application of DMAIC becomes justifiable.

However, there are also many simpler single-issue problems, usually managed by a Green Belt, that fall mostly within the boundary of a department. The confidence level of the solutions may also be less than 95%. Furthermore, many service organisations and SMEs may gain by using the costs and time of employing DMAIC in full against the benefits and costs of skipping some steps to find a solution. Thus, DMAIC Lite is appropriate for these smaller projects.

3.3.2 Characteristics of DMAIC Lite

The following characteristics of DMAIC Lite are identified:

- A part-time team led by a Green Belt works for a relatively short period (usually less than three months).
- The 'Define' step is carried out in detail to clearly establish the project charter.
- The 'Measure' and 'Analyse' phases are carried out together.
- The 'Improve' and 'Control' steps are executed concurrently.
- The problem is usually with a single issue encompassed in one section or two to three departments.
- Bias for action is 80–90% confident as compared with over 95% in a typical DMAIC process.
- Implementation is less risky with a modest investment.

3.3.3 When to Use DMAIC Lite

- When the above characteristics are appropriate.
- When, to get to the root of the problem, you do not need to bring together people of different experience and expertise.
- When implementation risk is not high and usually confined to one department.
- When the investment in DMAIC training for Black Belts is too expensive, especially for SMEs.

3.3.4 How to Conduct the DMAIC Lite Process

Similar to the DMAIC process, the DMAIC Lite procedure has five structured steps – although the boundaries between 'Measure' and 'Analyse' and also between 'Improve' and 'Control' are more flexible in the DMAIC Lite version. The steps involved are briefly described below.

3.3.4.1 Define

The purpose of 'Define' is to enable both the team and the sponsor to reach agreement on the scope, group composition, timeline and performance targets for the project. The key deliverables or outputs are:

- A completed project charter.
- VOC (both internal and external).
- A SIPOC diagram, including a high-level process map.
- Completed project plans, including Gantt chart, stakeholder analysis and force field analysis.
- A rough estimate of costs and benefits.

The Define stage should be signed off by the sponsor by a completed Define 'gate review'. If you have to spend more than one week (part time) in Define, this could be a warning sign worth heeding. It may be an indication that the scope is too broad or the problem is more complex than anticipated. In this case, the sponsor may then decide to switch to a full DMAIC process led by a Black Belt.

3.3.4.2 Measure and Analyse

In DMAIC Lite, Measure and Analyse are carried out as a combined step and the gate review is performed after completing the Analyse phase. The purpose of this stage is to understand the problem, collect reliable data and pinpoint the root causes affecting the variables of the project goals.

The key deliverables or outputs are:

- Revised value stream and process maps and detection of bottlenecks.
- A revised project charter.
- Reliable data on critical inputs (Xs) and outputs (Ys) to be used to analyse the variables (e.g. defects, speed, throughputs) of the project goals.
- Identification of non-value-added and value-added activities.
- Data charts and analyses to show the link between input and output of variables and the critical Xs as potential root causes.
- An updated estimate of costs and benefits.

This merged stage is signed off by the sponsor after the completion of Analyse. The gate review should focus on process analysis, root cause analysis and an updated project charter. If a quick hit or 'low-hanging fruit' opportunity is ascertained then the sponsor may recommend implementation following a Kaizen Event approach.

3.3.4.3 Improve and Control

In DMAIC Lite, Improve and Control are also executed as a combined step and the gate review is carried out after completing Control when the project is closed. The purpose of this stage is to implement the solution, conclude the project and hand over the improved process to the process owner.

The key deliverables or outputs are:

- An improved process that is stable, predictable and meets the project goals.
- Before and after data on metrics related to the process showing the indication of projected benefits.
- A system for monitoring the metrics of the implemented process.
- A document showing the updated operating procedure and training needs.
- A short project report.

This combined stage is signed off by the sponsor after the completion of Control and the closure of the project. The gate review should focus on implementation results, learning points, measures and documented processes for ongoing operation and the acceptance by the process owner.

3.3.5 Training Requirements of DMAIC Lite

The team leaders of projects using DMAIC Lite are expected to be trained at least up to a Green Belt standard with a broad understanding of FIT SIGMA and the DMAIC process. The workshop for Green Belt training usually extends for a week and includes the learning of basic FIT SIGMA tools. The participants of the project team should also acquire an awareness of DMAIC in one- or two-day workshops.

3.4 KAIZEN EVENT

3.4.1 Background

A Kaizen Event is a structured process improvement approach comprising a small group of people to improve a specific aspect of the business process in a rapid and focused manner. This method is applicable to all types and sizes of organisations and is usually employed when the root causes of the problem are known. It should be noted that there are different names and types of Kaizen Events which may cause some confusion. For reference, some of these terms are Rapid Improvement Kaizen, Flow Kaizen, Point Kaizen, Sigma Kaizen and the Kaizen Blitz. The idea behind the Kaizen Blitz is to assemble the team in an environment like a war room to plan and implement the changes very quickly. It has evolved in the application of fast-track Lean methods and it is also being adapted to a flexible DMAIC framework.

The structured approach of a Kaizen Event is to follow a methodology of SCORE (Select, Clarify, Organise, Run and Evaluate) which bears a resemblance to DMAIC (Define, Measure, Analyse, Improve and Control). However, the main difference between SCORE and DMAIC is that SCORE is less rigorous, less data-driven and less time-consuming. Another attraction of a Kaizen Event following SCORE is that it does not require a formal Black Belt or Green Belt type training programme, although some education is required for a winning Kaizen Event. The success of a Kaizen Event also demands attention to detail, a participative team and a mind set for change management.

3.4.2 Characteristics of a Kaizen Event

George *et al.* (2005) identify the following characteristics of a Kaizen event:

- A dedicated team works full time for a short period (usually a week).
- Most of the basic data and scope of the project is already available.
- Management makes support facilities (e.g. 'war room', flip charts, information technology, maintenance, etc.) available.
- Bias for action is 70–80% confident as compared with over 95% in a typical DMAIC process.
- Implementation is immediate (usually within three weeks).

3.4.3 When to Use a Kaizen Event

- When the above characteristics are appropriate.
- When obvious waste is identified or a problem is clearly defined.
- When implementation risk is minimal.
- When the investment in DMAIC training is doubtful, especially for SMEs.

3.4.4 How to Conduct the SCORE Process in a Kaizen Event

Similar to the DMAIC process, the SCORE procedure consists of five structured steps, although the boundaries between each phase are more flexible in SCORE. It is expected that each stage is accomplished within one day. These steps are briefly described below:

3.4.4.1 Select

The major activities in Select should include:

- Select a Kaizen leader and participants and notify them (carried out by the sponsor).
- Select the problem with a predefined project brief and the relevant data that are already available (usually defined by a Black Belt or the sponsor).
- Organise the 'war room' with relevant facilities and logistics (e.g. flip charts, stationery, computers, refreshments, meals, etc.) (conducted by the team leader).
- Arrange for coverage during participants' absence from the workplace (responsibility of the participants).
- Alert departments and key stakeholders whose support is likely to be required during the week (by team leader).
- Arrange for sponsor participation (again done by team leader).

3.4.4.2 Clarify

The major activities in Clarify should include:

- Validate the value stream and process maps of the problem (by participants).
- Verify the project objectives and available data provided (by team leader).
- Decide if additional data is required (by team leader).
- Collect and measure additional data if necessary (by participants).

3.4.4.3 Organise

The major activities in Organise should include:

- Validate root causes with a cause and effect diagram and pinpoint the sources of waste.
- Review waste reduction techniques (e.g. SMED, five Ss, new layout, new method, etc.).
- Brainstorm process improvements by eliminating non-value-added tasks.

3.4.4.4 Run

The major activities in Run should include:

- Create an action plan.
- Consult department managers affected by the changes.
- Test the new process and fine tune.
- Ensure that the process is stable.

3.4.4.5 Evaluate

The major activities in Evaluate should include:

- Train employees affected by the changes.
- Create standard operating procedures to document the changes.
- Report results and present them to the management team.
- Establish a plan to monitor results for continuous improvement.

3.4.5 Training Needs for a Kaizen Event

The success of a Kaizen Event is underpinned by the application of a structured SCORE process as described above. Therefore, it is important that the team leader and some of the participants are trained in the basic tools of FIT SIGMA and the SCORE process. The team leader is expected to have an education to Green Belt standards and instruction in change management culture. It is also advisable to have an experienced Kaizen expert to facilitate the event and act as a mentor for the team leader. In addition, it is an effective practice to deploy an external Kaizen consultant to act as a trainer, facilitator and catalyst for change until the in-house capability of managing Kaizen Events is established.

3.5 SUMMARY

When we consider DMAIC and Kaizen, by their very different natures there are advantages and disadvantages to each approach. DMAIC Full projects may sometimes take longer than necessary, while Kaizen Events often do not bring the rigour of data-driven measurement and analysis of the solution. DMAIC Lite offers the rigidity in a shorter time scale but may also bear the risk of ‘cutting corners’. Therefore, the key to using the appropriate methodology in FIT SIGMA is to gain its strength and weakness by appropriate training. Secondly, it is vital to apply it with a good level of understanding for the right type of performance improvement opportunity. In a FIT SIGMA programme it is also feasible to integrate aspects of Kaizen Events to accelerate the DMAIC process.

There are scenarios when the three methods discussed in this chapter just may not be adequate. For example, when you are designing a process from scratch or you are devising a new product or service, design for Six Sigma (DFSS) should be more appropriate. DFSS is not covered in this book but can be found in *Implementing Six Sigma and Lean* (Basu, 2009).

4
FIT SIGMA Tools

4.1 INTRODUCTION

The selection and application of appropriate tools is a critical success factor of a FIT SIGMA programme. There are many tools and techniques available in the Six Sigma world (Basu, 2009), some of which are more advanced – such as design of experiments (DOE), quality function deployment (QFD), failure mode and effect analysis (FMEA) and statistical process control (SPC). In this chapter, we do not introduce any 'new' 'FIT Σ tools', however, those discussed do constitute the more frequently and appropriately used Six Sigma and Lean tools. Thomas Edison once said, 'Your idea has to be original only in its adaptation to the problem you are currently working on'. The adaptation of the existing tools represents 'appropriate FIT SIGMA tools'. These tools are described here as they are used more predominantly under the five steps of the DMAIC cycle.

4.2 TOOLS FOR 'DEFINE'

The tools for Define as described in this section are:

- value stream mapping;
- SIPOC diagram;
- flow diagram;
- CTQ tree;
- project charter.

4.2.1 D1: Value Stream Mapping

4.2.1.1 Definition

Value stream mapping (VSM) is a visual illustration of all activities required to bring a product through the main flow, from raw material to the stage of reaching the customer.

Mapping out the activities in a production process with cycle times, down times, in-process inventory and information flow paths helps us to visualise the current state of the process and guides us to the future improved state.

4.2.1.2 Application

VSM is an essential tool of Lean Manufacturing in identifying non-value-added activities at a high level of the total process.

According to Womack and Jones (1998), the initial objective of creating a value stream map is to identify every action required to make a specific product. Thus the first step is to group these activities into three categories:

- Those which actually create value for the customer.
- Those which do not create value but are currently necessary (type one 'muda').
- Those which create no value as perceived by the customer (type two 'muda').

Once the third set has been eliminated, attention is focused on the remaining non-value-creating activities. This is achieved through making the value flow at the pull of the customer.

VSM is closely linked with the analytical tool of process mapping. Having established improvement opportunities at a high level by VSM, a detailed analysis of the specific areas of the process is effective with process mapping.

4.2.1.3 Basic Steps

1. The first step of VSM is to select the product or process for improvement.
2. Each component of production from the source to the point of delivery is then identified.
3. The entire supply chain of the product or process (e.g. through order entry, purchasing, manufacturing, packaging and shipping) is mapped sequentially.
4. The quantitative data of each activity (e.g. storage time, delay, distance travelled, process time and process rate) is then recorded.
5. Each component (i.e. activity) of the production or process is evaluated to determine the extent to which it adds value to product quality and production efficiency.
6. These activities are then categorised as:
   - value-added;
   - necessary non-value-added;
   - unnecessary non-value-added.
7. Areas of further analysis and improvement are then identified clearly.

4.2.1.4 Worked-out Example

The following example is adapted from Womack and Jones (1998, pp. 38–43).

Consider a case containing eight cans of cola at a Tesco store.

Figure 4.1 shows a value stream map of cola, from the mining of bauxite (the source of the aluminium in the cans) to the user's home.

The quantitative data related to the activities in the value stream are summarised in Table 4.1.

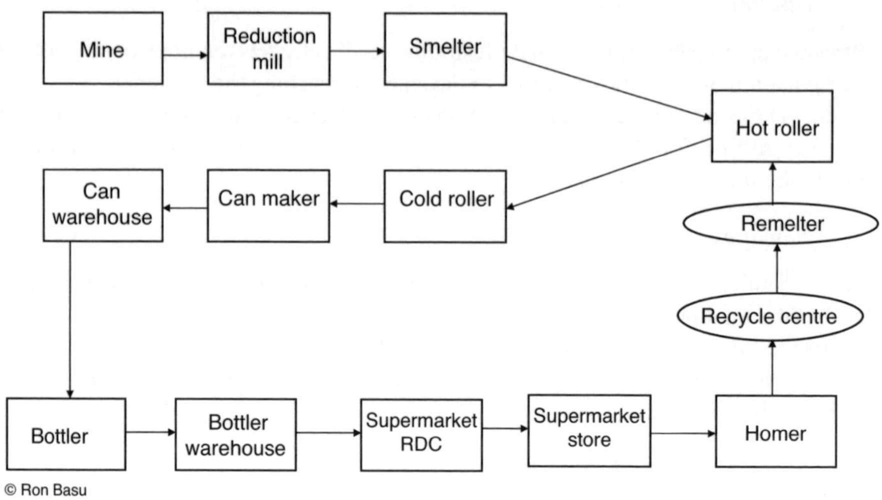

Figure 4.1 Value stream for cola cans

Table 4.1 Quantitative data of cola cans

	Incoming storage	Process time	Finished storage	Process rate	Cumulative days
Mine	0	20 min	2 weeks	1000 t/hr	319
Reduction mill	2 weeks	30 min	2 weeks	—	305
Smelter	3 months	2 hours	2 weeks	—	277
Hot rolling mill	2 weeks	1 min	4 weeks	10 ft/min	173
Cold rolling mill	2 weeks	< 1 min	4 weeks	2100 ft/min	131
Can maker	2 weeks	1 min	4 weeks	2000/min	89
Bottler	4 days	1 min	5 weeks	1500 min	47
Supermarket RDC	0	0	3 days	—	8
Supermarket store	0	0	2 days	—	5
Home storage	3 days	5 min	—	—	3
Totals	**5 months**	**3 hours**	**6 months**	**—**	**319**

It is evident from the details in Table 4.1 that value-added activities take only three hours compared with the total time (319 days) from the mine to the recycling bin. This proportion is surprisingly small when one considers the overall duration of the process.

4.2.1.5 Training Requirements

The basic principles of VSM are not new, but making sense of these ideas and applying them to practical problems clearly requires some tuition. There is no shortage of training consultants offering workshops and courses in the tools of Lean Manufacturing, including VSM.

We recommend that the team members should undergo an education workshop of at least half a day's duration for VSM. This training programme should be combined with other relevant tools, like process mapping.

4.2.2 D2: SIPOC Diagram

4.2.2.1 Definition

SIPOC is a high-level map of a process to view how a company goes about satisfying a particular customer requirement in the overall supply chain. SIPOC stands for:

Supplier – the person or company that provides the input to the process (e.g. raw materials, labour, machinery, information, etc.). The supplier may be both external and internal to the company.

Input – the materials, labour, machinery, information, etc. required for the process.

Process – the internal steps necessary to transform the input to output.

Output – the product (both goods and services) being delivered to the customer.

Customer – the receiver of the product. The customer could be the next step of the process or a person or organisation.

4.2.2.2 Application

A SIPOC diagram is usually applied during the data collection of a project or at the 'Define' stage of DMAIC in a Six Sigma programme. However, its impact is utilised throughout the project life cycle.

SIPOC not only shows the inter-relationships of the elements in a supply chain, but also any critical to quality indicators (CTQ) such as 'delivered in seven days'.

4.2.2.3 Basic Steps

1. Select the process for the SIPOC diagram and identify CTQ parameters.
2. Determine the input requirements.
3. Identify the suppliers for each of the input elements.
4. Define the output and validate CTQ parameters.
5. Identify the customers.
6. Draw the SIPOC process diagram.
7. Retain the diagram for the rest of the improvement project.

4.2.2.4 Worked-out Example

Figure 4.2 shows a SIPOC diagram for a company that leases equipment.

Supplier	Input	Process	Output	Customer
Credit Agency	• Credit report • Response in 30 minutes • Review report in 3 minutes • Lease schedule • Retail price data	Customer credit review Equipment validation Preparation of document	Lease agreement • Complete in 7 days • Payment in 5 days • Maintenance included	Equipment lessor
Engineering Department	• Equipment • Fork lift truck 1500 kg	Funding approved		

Source: Adapted from George (2002, p. 185).

Figure 4.2 SIPOC process diagram

4.2.2.5 *Training Requirements*

The preparation and teaching of SIPOC should be amalgamated with those for the CTQ tree and the IPO diagram since they complement one another. The combined training is expected to be carried out for both Black Belts and Green Belts in half a day.

4.2.3 D3: Flow Diagram

4.2.3.1 *Definition*

A flow diagram (also called a flow chart) is a visual representation of all major steps in a process. It helps a team to understand a process better by identifying the actual flow or sequence of events in a process that any product or service follows.

There are variations to a flow diagram depending on the details required in an application. We have included two other forms of illustration in this family. These are the flow process chart (see Chapter 5) and process mapping (see Chapter 6).

The type of flow diagram described in this section is a top-level mapping of the general process flow and uses four standard symbols:

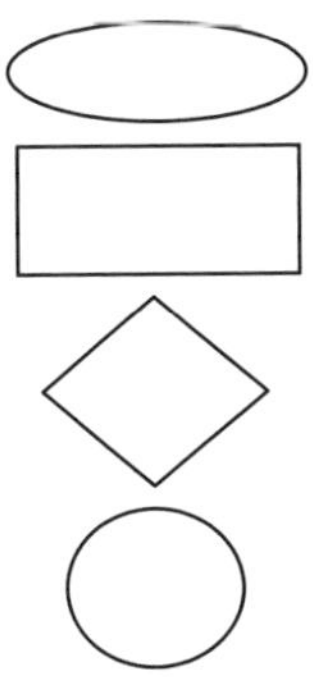

- An oval is used to show the start and end of a process.
- A box or a rectangle is used to show a task or activity performed in the process.
- A diamond shows those points in the process where a decision is required.
- A circle with either a letter or a number identifies a break and connects to another page or part of the diagram.

4.2.3.2 *Application*

A flow diagram can be applied to any type of process. It can be used for anything from the development of a product to the steps involved in making a sale or servicing a product.

It allows a team to come to a consensus regarding the steps of the process and to identify critical and problem areas for improvement.

In addition, it serves as an excellent training aid to understand the complete progression of the course of action.

4.2.3.3 *Basic Steps*

1. Select the process and determine the scope or boundaries of that process:
   (a) Clearly define where the process understudy starts and ends.
   (b) Agree the level of detail to be shown on the flow diagram.
2. Brainstorm a list of major activities and determine the steps in the process.
3. Arrange the steps in the order they are carried out. Unless you are developing a new process, sequence what actually is and not what should be.
4. Draw the flow diagram using the appropriate symbols.

5. There are a number of good practices when charting a process, including:
   (a) Use Post-It™ notes so that you can move them around in a large process.
   (b) For a large-scale process, start by charting only the major steps or activities.
   (c) Come back to develop further details for major steps if necessary.
   (d) Consider process mapping (see Chapter 6) to apply to a larger process.

4.2.3.4 Worked-out Example

Figure 4.3 shows an example of a flow diagram to illustrate an advertising process.

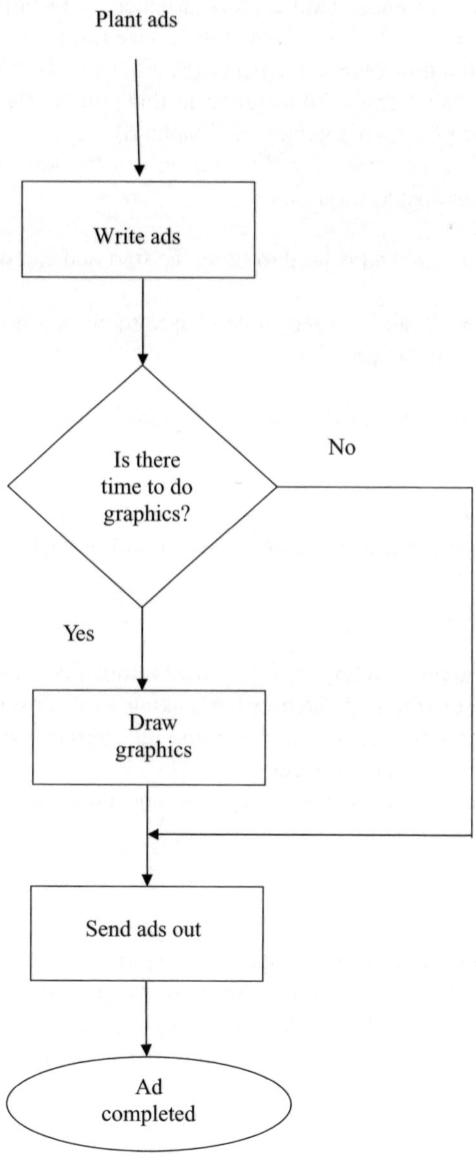

Figure 4.3 A flow diagram

4.2.3.5 Training Requirements

The basic principles of a flow diagram are very simple and can be acquired in a self-teaching process by the user. However, good experience is required to develop a flow diagram of a more complex process. This sort of knowledge can only be gained by working as a team involved with improvement projects.

4.2.4 D4: CTQ Tree

4.2.4.1 Definition

'Critical to quality' (CTQ) is a term widely used within the field of Six Sigma activities to describe the key output characteristics of a process. An example may be an element of a design or an attribute of a service that is critical in the eyes of the customer.

A CTQ tree helps the team to derive the more specific behavioural requirements of the customer from his general needs.

4.2.4.2 Application

A CTQ tree is a useful tool during the data collection stage (Define) of an improvement project. Once the project team has established who their customers are, the working party should then move towards determining the customer needs and requirements. The 'needs' of a customer form the output of a process. 'Requirements' are the characteristics necessary to determine whether the customer is happy with the output delivered. These constitute what is 'critical to quality', and thus a CTQ tree helps to identify these CTQs in a systematic way.

4.2.4.3 Basic Steps

1. Identify the customer.
2. Pinpoint customers' general needs in Level 1.
3. Distinguish the first set of requirements for that need in Level 2.
4. Drill down to Level 3 if necessary to ascertain the specific behavioural requirements of the customer.
5. Validate the requirements with the customer. The process of validation could be one-to-one interviews, surveys or focus groups, depending on the CTQ.

4.2.4.4 Worked-out Example

The following example is adapted from Eckes (2001, p. 55).

Figure 4.4 shows a worked-out example of a CTQ tree for room service in a hotel.

4.2.4.5 Training Requirements

A CTQ tree is a simple tool in principle and a one-hour training session (including an exercise) should be sufficient for a team member to get involved in developing a practical CTQ tree. This is appropriate for both Black Belts and Green Belts.

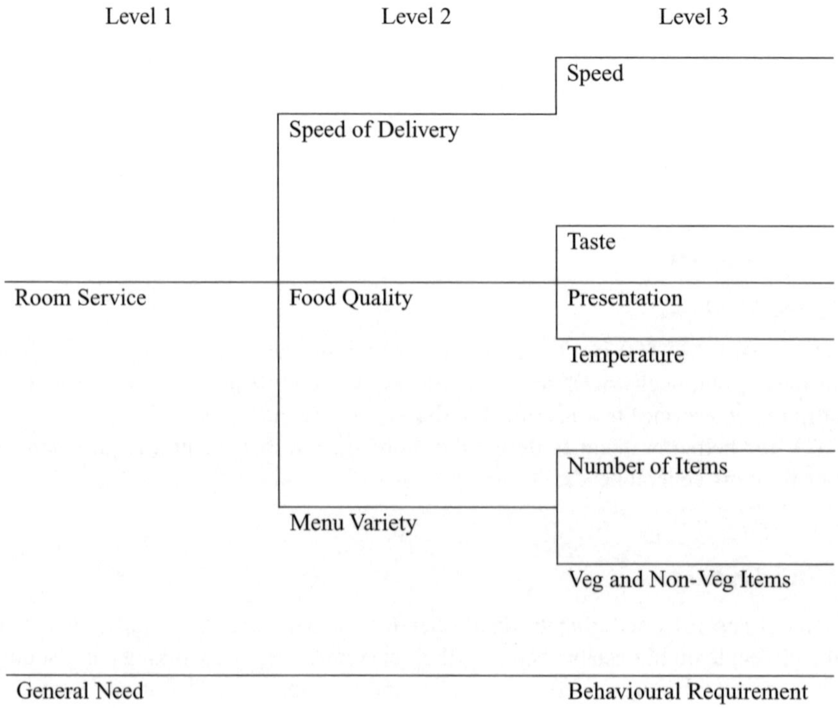

Figure 4.4 A CTQ tree

4.2.4.6 Final Thoughts

A CTQ tree is a simple but powerful tool for capturing the details of customer requirements and we recommend its use at the very early stage of a Six Sigma project.

4.2.5 D5: Project Charter

4.2.5.1 Definition

A project charter is a working document for defining the terms of reference of each Six Sigma project. The charter can make a successful project by specifying necessary resources and boundaries that will in turn ensure success.

The necessary elements of a project charter include:

- Project title – it is important to use a descriptive title that will allow others to quickly identify the project.
- Project type – whether it is for quality improvement, increasing revenue or reducing cost.
- Project description – a clear explanation of the problem, the opportunity and the goal.
- Project purpose – why you are carrying out this project.
- Project scope – project dimensions, what is included and not included.
- Project objectives – target performance improvement in measurable terms.
- Project team – sponsor, team leader and team members. It is important to identify Black Belts and Green Belts within the team.

- Customers and CTQs – both internal and external customers and CTQs specific to each customer.
- Cost benefits – a draft business case of savings expected and the cost required to complete the project.
- Timing – anticipated project start date and end date. Target completion dates of each phase (e.g. DMAIC) of the project are also useful.

4.2.5.2 Application

A project charter is the formalised starting point in the Six Sigma methodology. It takes place in the Define stage of DMAIC.

We recommend that larger (e.g. over $1m savings) projects of a Six Sigma programme, which are usually led by a Black Belt, should have a project charter. In fact, a project charter can also be useful in a multi-discipline, medium-sized project.

A project charter should be used as a working document that is updated as the project evolves. The version control of the charter is therefore very important.

4.2.5.3 Basic Steps

1. Select the project by taking into account the following criteria:
   (a) Not capital intensive.
   (b) Achievable in six months.
   (c) High probability of success.
   (d) Good fit with Six Sigma techniques.
   (e) Clearly linked to real business need.
   (f) Historical and current data accessible.
2. Identify the customers and their specific CTQ requirements by a SIPOC diagram.
3. Estimate order of magnitude figures for the costs and savings for the project. (These estimates would be updated when more accurate data becomes available as the project advances.)
4. Obtain top management support and sponsorship.
5. Select the project team with clear leadership and ownership for delivery.
6. Develop the project charter following a defined template (see Figure 4.5).
7. Obtain the approval of the sponsor.
8. Review and update the charter with the necessary version control as the task progresses.

4.2.5.4 Worked-out Example

Figure 4.5 shows an example of a project charter for a 'Safety Performance' scheme used by DuPont Teijin Films Ltd, UK.

4.2.5.5 Training Requirements

The training of a project charter is included in the education programmes for both Black Belt and Green Belt attainment. The basic principles of a project charter are relatively simple and

Project:	Safety Performance Predictor Model
Date:	3/28/00
Project Type:	Quality X Revenue Cost reduction
Problem Statement:	Plant needs early warning signals so programmes can be put in place to prevent injuries/incidents
Goal Statement:	Predict injuries/incidents prior to occurring Performance Level: 1.47 DPMO Time frame: 6 months
Project Scope and Approach:	Study safety data to determine what factors influence injuries and incidents. Use the most significant factors and develop a predictor model to serve as early warning signals for a potential deterioration in our safety performance.
Team Members:	Terry Leadership John Debbie Leader: Vickie Sponsor: Terry Mentor: Tim
Customers:	Leadership Terry Network Leaders
CTQs:	Early warning signals of downward shifts in safety climate Combined data source for all site safety data To be able to identify problem areas before injuries occur
Defect Definition:	Injuries and incidents
Opportunities per unit:	Exposure hours
Approx. DPMO:	1.47 Z st: 6.19
Goal DPMO:	1.25 Z st goal: 6.19
Stake:	Confidence interval: Upper: Lower capital:
Timing:	Define Measure Analyse Improve Control
Target Completion:	04/30/00 06/30/00 07/31/00 08/31/00 09/30/00
Data Issues:	Safety Climate Indicator measurements system needs to be validated

Figure 4.5 Six Sigma project charter

can be explained to team members in less than an hour. The development of a chapter with appropriate data may need a couple of days.

4.3 TOOLS FOR 'MEASURE'

The tools for Measure as described in this section are:

- run chart;
- histogram;
- cause and effect diagram;
- Pareto chart;
- control charts.

Table 4.2 Operational efficiency of a packaging machine

x	Jan	Feb	Mar	Apr	May	Jun	Jul	Aug	Sept	Oct	Nov	Dec
y	55	60	45	40	65	60	65	30	60	65	60	55

4.3.1 M1: Run Charts

4.3.1.1 Definition

A run chart is a graphical tool to allow a team to study observed data for trends over a specified period of time. It is basically a simple line graph of *x*- and *y*-axes.

4.3.1.2 Application

A run chart has a wide range of applications to detect trends, variation or cycles. It allows a team to compare performances of a process before and after the implementation of the solution. The application areas include sales analysis, forecasting, performance reporting and seasonality breakdown.

4.3.1.3 Basic Steps

1. Select the parameter and time period for measurement.
2. Collect data (generally 10–20 data points) to identify meaningful trends:
   (a) *x*-axis for time or sequence cycle (horizontal).
   (b) *y*-axis for the variable parameter that you are measuring (vertical).
3. Plot the data in a line graph along the *x*- and *y*-axes.
4. Interpret the chart. If there are no obvious trends then calculate the average value of the data points and draw a horizontal line at this mean value.

4.3.1.4 Worked-out Example

Table 4.2 shows the operational efficiency (%) of a packaging machine.
The run chart for the above data is shown in Figure 4.6.

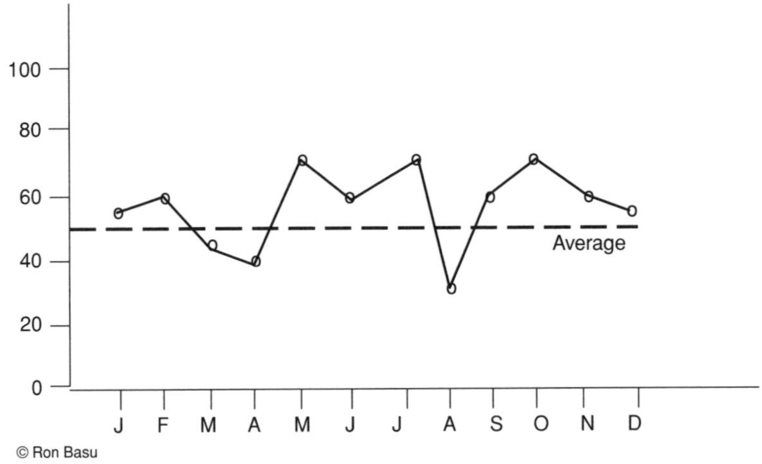

Figure 4.6 Example of a run chart

4.3.1.5 Training Requirement

The understanding and application of a run chart is so simple that it is not essential to conduct classroom training specifically for it. The team members should be able to use a run chart after a preliminary briefing, in a practical application. The training process involving a run chart is usually included in an education programme for continuous improvement tools. Due to the fact that it could be seen as encompassing simplicity to a fault, a danger in using a run chart is the inclination to interpret every variation as being noteworthy. For any statistically significant result, control charts or regression analysis should be more appropriate. We recommend the use of a run chart for detecting a visual trend only and identifying areas of further analysis.

4.3.2 M2: Histograms

4.3.2.1 Definition

A histogram is a graphical representation of recorded values in a data set according to the frequency of occurrence. It is a bar chart of numerical variables giving a graphical representation of how the data are distributed.

4.3.2.2 Application

The histogram is used extensively in both statistical analysis and data presentation. A histogram displays the distribution of data and thus reveals the amount of variation within a process. There are a number of theoretical models for various shapes of distribution, of which the most common one is the normal or Gaussian distribution.

There are several advantages of applying histograms in continuous improvement projects, including:

- They display large amounts of data that are difficult to interpret in tabular form.
- They illustrate quickly the underlying distribution data, revealing the central tendency and variability of a data set.

4.3.2.3 Basic Steps

1. Collect at least 50–125 data points for establishing a representative pattern.
2. Subtract the smallest individual value from the largest in the data set.
3. Divide this range by 5, 7, 9 or 11 depending on the number of data points. As a rough guide, take the square root of the total number of data points and round it to the nearest integer. For example, for 50 data points, divide the range by 7 and for 125 data points you should divide by 11.
4. The resultant value determines the interval of the sample. It should be rounded up for convenience.
5. Calculate the number of data points in each group or class.
6. Plot the histogram with the intervals in the x-axis and the frequency of occurrence on the y-axis.
7. Clearly label the histogram.
8. Interpret the histogram related to centring, variation and shape (distribution).

Table 4.3 Miles per gallon data of a car pool

18	16	30	29	28	21	17	41	8	17
32	26	16	24	27	17	17	33	19	18
31	27	23	38	33	14	13	26	11	28
21	19	25	22	17	12	21	21	25	26
23	20	22	19	21	14	45	15	24	34

4.3.2.4 Worked-out Example

The following example is taken from Schmidt *et al.* (1999, pp. 135–137).

Table 4.3 shows the data points of miles per gallon data of a car pool.

For a 50-point data set we would use $\sqrt{50} = 7.06 \rightarrow 7$ classes. Since our smallest data point is 8 and the largest is 45, the interval should be $(45 - 8)/7 = 5.3$. We round it up to make it 6. In Table 4.4 we place individual data in each class as shown.

Next we construct the histogram as shown in Figure 4.7. We see that the most values are between 12 and 30 mpg, with 18–24 mpg representing the median. The data are also skewed to the right by some high mileage figures.

4.3.2.5 Training Needs

Although a histogram can lead to complex statistical analysis, its basic principles are relatively simple. Most members of the project team are likely to have some familiarity with and knowledge of histograms. However, a one-hour session of classroom training should be valuable to establish the methodology, especially related to the range, class and intervals.

4.3.3 M3: Cause and Effect Diagrams

4.3.3.1 Definition

The cause and effect diagram is a graphical representation of potential causes for a given effect.

Since it was first used by Ishikawa, this type of illustration is also known as an Ishikawa diagram. In addition, it is often referred to as a 'fishbone' diagram due to its skeletal appearance.

The purpose of the illustration is to assist in brainstorming and enabling a team to identify and graphically display, in increasing detail, the root causes of a problem.

Table 4.4 Class placement of individual data

Class no.	Class range	Values	Frequency of occurrence
1	–6–12	8, 11	2
2	12–18	16, 16, 17, 17, 14, 12, 14, 17, 17, 13, 15, 17	12
3	18–24	18, 21, 23, 19, 20, 23, 22, 22, 19, 21, 21, 21, 21, 19, 18	15
4	24–30	26, 27, 25, 29, 24, 28, 27, 26, 25, 24, 28, 26	12
5	30–36	32, 31, 30, 33, 33, 34	6
6	36–42	38, 41	2
7	42–48	45	1

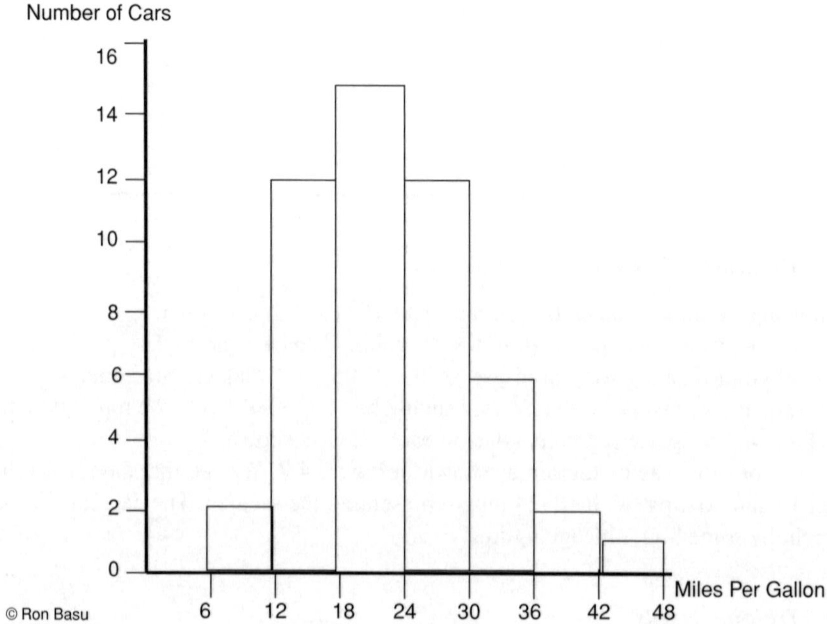

Figure 4.7 Example of a histogram

4.3.3.2 Application

The cause and effect diagram is arguably the most commonly used of all quality improvement tools. The 'effect' is a specific problem and is considered to constitute the head of the diagram. The potential causes and sub-causes of the problem form the bone structure of the skeletal fish.

They are typically used both during the measurement and analysis phase of the project. Their wide application area covers Six Sigma teams, TQM teams or continuous improvement teams as part of brainstorming exercises to identify the root causes of a problem and offer solutions. It focuses the group on causes, rather than symptoms.

There are a number of variants in the application of a cause and effect diagram. The two most common types are the 6M diagram and CEDAC. Let us examine each of these terms a little more closely.

In a 6M diagram, the main bone structure or branches typically consist of the self-explanatory '6Ms':

- Machine.
- Manpower.
- Material.
- Method.
- Measurement.
- Mother Nature (environment).

A CEDAC (cause and effect diagram assisted by cards) diagram works slightly differently. A blank, highly visible fishbone chart is displayed in a meeting room. Every member of the team must post both potential causes and solutions on a card (or Post-It™ note) considering each of the categories.

A CEDAC also consists of two major formats:

- dispersion analysis type;
- process classification type.

Firstly, the dispersion analysis type is used usually after 6M or CEDAC diagrams have been completed. The major causes identified are then treated as separate branches and their sub-causes are identified.

However, the process classification type uses the major steps of the process in place of the major cause categories. This form is usually used when the problem encountered cannot be isolated to a single department. Each stage of the process is then analysed by using a 6M or CEDAC approach.

A typical sequence of cause and effect diagrams for a complex problem could be: process classification type; 6M or CEDAC; dispersion analysis type.

4.3.3.3 Basic Steps

1. Select the most appropriate cause and effect format. If the problem can be isolated to a single section or department, choose either a 6M (small-team) or a CEDAC (large-team) approach.
2. Define with clarity and write the key effect of the problem in a box to the right-hand side of the diagram.
3. Draw a horizontal line from the left-hand side of the box. Draw main branches (fish bones) of the diagram after agreeing the major categories (e.g. 6M) of causes.
4. Brainstorm for each category the potential sub-causes affecting the category.
5. List the sub-causes of each category in a flipchart. In a CEDAC approach these would be a collection of Post-It™ notes for each section.
6. Rank the sub-causes in order of importance by a group consensus (or multi-voting) and select up to six top sub-causes for each category.
7. Construct the diagram by posting the top sub-causes in each category. These are the 'root causes'.
8. Decide upon further dispersion analysis or gather additional data needed to confirm the root causes.
9. Develop solutions and improvement plans.

4.3.3.4 Worked-out Example

The following example is taken from Basu and Wright (2003, pp. 29–30).

Consider the situation where customers of a large international travel agency sometimes find that when they arrive at their destination, the hotel has no knowledge of their booking.

In this case, to get started you might begin with 'hotel not booked' as the effect and the 6Ms as the causes. When the sub-causes are further investigated, the diagram may look like that shown in Figure 4.8.

The diagram points out that many sub-causes – including training, e-mail systems and customer feedback – appear to be worthwhile areas to follow up.

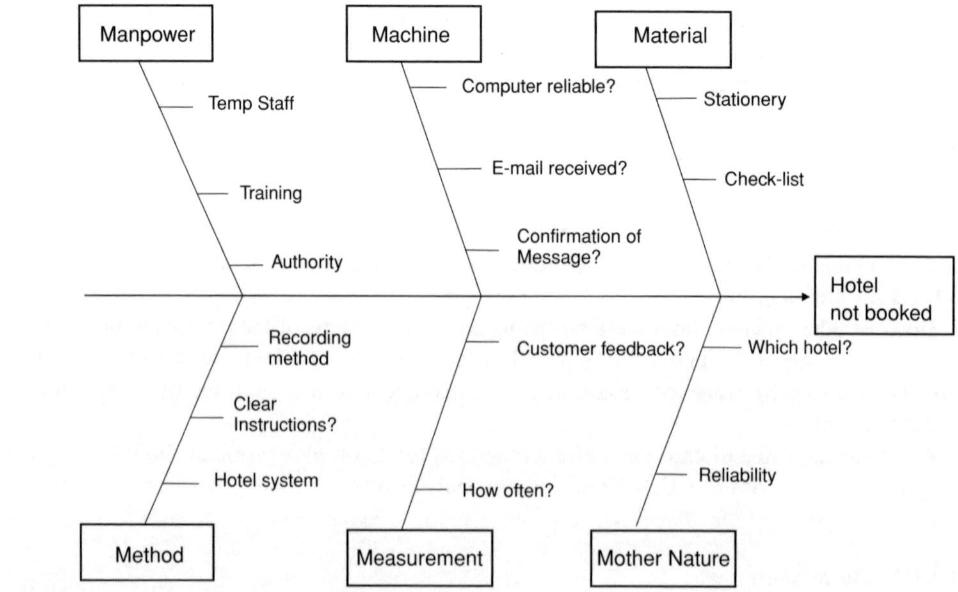

Figure 4.8 Cause and effect diagram

4.3.3.5 Training Requirements

As cause and effect diagrams are essential to a quality improvement programme, it is important that members of the team receive at least one hour of hands-on training. An effective method of instruction is by participating in a brainstorming exercise, where a cause and effect diagram is developed. It is more important to identify the root causes than to assign the cause to a specific category. For instance, in the example given in Figure 4.8, 'clear instructions' could also be grouped under Manpower while 'checklist' could be assigned to the category of Measurement.

4.3.4 M4: Pareto Charts

4.3.4.1 Definition

A Pareto chart is a special form of bar chart that rank-orders the bars from highest to lowest in order to prioritise problems of any nature.

It is known as 'Pareto' after a nineteenth-century Italian economist Wilfredo Pareto who observed that 80% of the effects are caused by 20% of the causes: 'the 80/20 rule'.

4.3.4.2 Application

Pareto charts are applied to analyse the priorities of problems of all types, e.g. sales, production, stock, defects, sickness, accident occurrences, etc. Improvement efforts are directed to priority areas that will have the greatest impact.

There are usually two variants in the application of Pareto charts. The first type is the standard chart, where bar charts are presented in descending order. The second category is also known as the 'ABC analysis', where:

- Cumulative percentage values of causes are plotted along the x-axis.
- Cumulative percentage values of effects are plotted along the y-axis.
- 80% of the effects with the corresponding percentage of causes are grouped as category 'A'.
- 80–96% of the effects and the corresponding causes are 'B' items.
- The remaining values are category 'C'.

4.3.4.3 Basic Steps

The following steps apply for the preparation of a Pareto chart:

1. Identify the general problem (e.g. IC board defects) and its causes (e.g. soldering, etching, moulding, cracking, other).
2. Select a standard unit of measurement (e.g. frequency of defects or money loss) for a chosen time period.
3. Collect data for each of the causes in terms of the chosen unit of measurement.
4. Plot the Pareto chart with causes along the x-axis and the unit of measurement along the y-axis. The causes are charted in descending order of values from left to right.
5. Analyse the graph and decide on the priority for improvement.

The following steps apply for the preparation of an ABC analysis:

1. Decide on the causes (e.g. number of customers) and effect (e.g. sales) of the problem areas that you want to prioritise.
2. Select a standard unit of measuring the effect (e.g. $ for sales values).
3. Collect data for each cause (e.g. customer) and the corresponding effect (e.g. sales in $).
4. Rank the cause according to the value of the effects (e.g. customers in descending order of $ sales).
5. Calculate the cumulative percentage values of the causes and effects.
6. Plot the cumulative percentage values of the causes (e.g. number of customers) along the x-axis and the cumulative percentage values of effects (e.g. sales in $) along the y-axis.
7. Identify A, B and C categories (e.g. A for 80% of sales, B for 80–96% of sales and C for the remainder).
8. Analyse the graph and decide on the priority for improvement.

4.3.4.4 Worked-out Examples

The following example of a Pareto chart is taken from Schmidt *et al.* (1999, p. 144).

The major defects identified during the manufacture of integrated circuit boards for a given period were as listed in Table 4.5.

We plot the causes along the x-axis and frequencies along the y-axis as shown in Figure 4.9.

An example of ABC analysis for a set of inventory items is shown in Figure 4.10.

4.3.4.5 Training Needs

The basic principles of a Pareto chart and ABC analysis are easy to follow. The training for this prioritising tool is usually included in the classroom education sessions related to the tools for measurement. A member of a team can be adept in the application of Pareto charts after, say, one hour's practice on a few practical problems.

Table 4.5 Defects identified in IC boards

Causes	Frequency	% Frequency
Soldering	60	40
Etching	40	27
Moulding	30	20
Cracking	15	10
Other	5	3
Total	**150**	**100**

4.3.5 M5: Control Charts

4.3.5.1 Definition

A control chart consists of a graph with time on the horizontal axis and an individual measurement (such as mean, or range) on the vertical axis.

A control chart is a basic graphical tool of statistical process control (see Chapter 9) for determining whether a process is stable and also for distinguishing usual (or common) variability from unusual (special assignable) causes. Three control limits are drawn: the central line (CL), the lower control limit (LCL) and the upper control limit (UCL). The points above the UCL or below the LCL indicate a special cause. If no signals occur then the process is assumed to be under control, i.e. only common causes of variation are present.

4.3.5.2 Application

Control charts can be used for examining a historical set of data and also for contemporary information. The current control based on current data underpins a feedback control loop in the process.

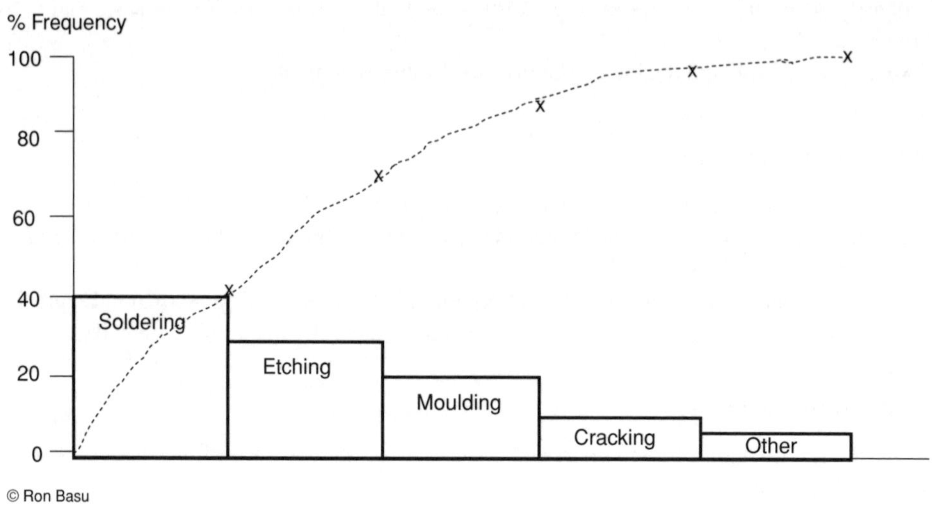

Figure 4.9 Pareto chart

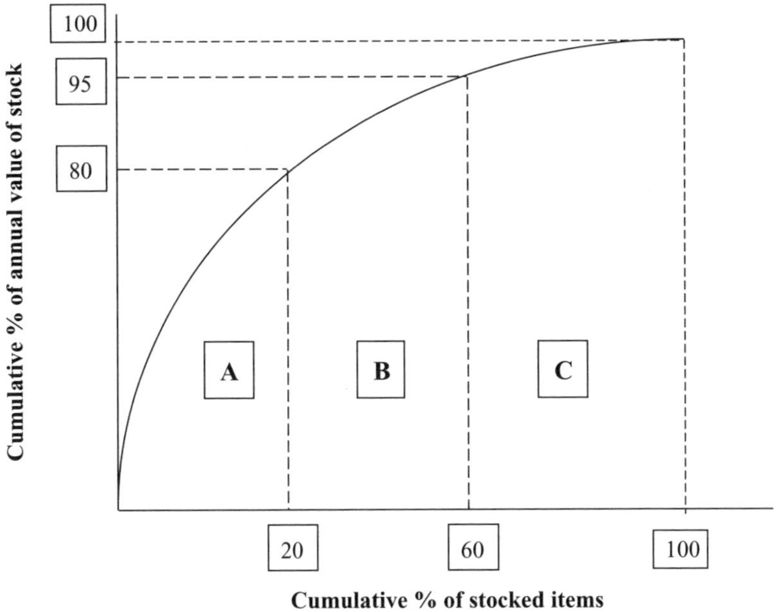

Figure 4.10 ABC analysis

There are many good reasons why control charts have been applied successfully in both quality control and improvement initiatives since Walter Shewhart first introduced the concept at Bell Laboratories in the early 1930s.

First, control charts establish what is to be controlled and force their resolution. Second, they focus attention on the process rather than on the product. For example, a poor product can result from an operator error, but a poor production process is not capable of meeting standards on a consistent basis. The third factor is that they comprise a set of prescribed techniques that can be applied by people with appropriate training in a specified manner.

For many probability distributions, most of the probability is within three standard deviations of the mean. So μ and σ are respectively the stable process mean and standard deviation, then

$$\mathrm{CL} = \mu$$
$$\mathrm{UCL} = \mu + 3\sigma$$
$$\mathrm{LCL} = \mu - 3\sigma$$

When the mean (μ_s) and standard deviation (σ_s) values are calculated from a sample of n, then

$$\mu_s = \mu$$
$$\sigma_s = \sigma/\sqrt{n}$$

There are two types of control charts. A variable chart is used to measure individual measurable characteristics, whereas an attributes chart is used for go/no-go types of inspection.

The x-bar chart (also called the mean chart), the s-chart (also called the standard deviation chart) and the R-chart (also called the range chart) are used to monitor continuous measurement or variable data.

Table 4.6 Packing carton data

Reading no.	Measurements				Average	Std dev.	Range
1	25.1	25.5	25.0	25.1	25.175	0.222	0.50
2	24.8	25.2	25.1	24.9	25.000	0.183	0.40
3	25.1	25.2	25.2	25.2	25.175	0.050	0.10
4	25.1	25.4	24.8	25.0	25.075	0.250	0.60
5	25.2	24.7	24.9	25.3	25.025	0.275	0.60
6	25.2	25.2	25.0	25.1	25.125	0.096	0.20
7	25.2	25.2	25.2	25.3	25.225	0.050	0.10
8	25.2	25.1	25.3	25.0	25.150	0.129	0.30
9	24.9	25.1	25.2	24.8	25.000	0.183	0.40
10	25.1	25.1	25.3	25.4	25.225	0.150	0.30
Average					**25.118**	**0.159**	**0.35**

The stable control charts are used to determine process capability, that is whether a process is capable of meeting established customer requirements or specifications.

4.3.5.3 Basic Steps

1. Choose the quality characteristic to be charted. A Pareto analysis is useful to identify a characteristic that is currently experiencing a high number of non-conformities.
2. Establish the type of control chart to ascertain whether it is a variable chart or an attribute chart.
3. Choose the sub-group or sample size. For variable charts, samples of four to five are sufficient, whereas for attribute charts samples of 50 to 100 are often used.
4. Decide on a system of collecting data. The automatic recording of information by a calibrated instrument is preferable to manually recorded data.
5. Calculate the mean and standard deviation and then determine the control limits.
6. Plot the data and control limits on a control chart and interpret the results.

4.3.5.4 Worked-out Example

The following example illustrates the construction of variable control charts based on the data of packing cartons for a morning shift. The information in Table 4.6 is taken from Ledolter and Burnill (1999).

Hence

$$\begin{aligned}
\text{CL} &= \mu_s = 25.118 \\
\sigma_s &= \sigma/\sqrt{n} = 0.159/\sqrt{10} = 0.159/3.163 = 0.05 \\
\text{UCL} &= \mu_s + 3\sigma_s = 25.118 + 3\,(0.05) = 25.27 \\
\text{LCL} &= \mu_s - 3\sigma_s = 25.118 - 3\,(0.05) = 24.97
\end{aligned}$$

The control limits and the data are plotted as shown in Figure 4.11.

Similar charts are drawn for the standard deviation and the range of the data. The control limits are constructed in such a manner that we expect approximately 99.7% of all data to fall

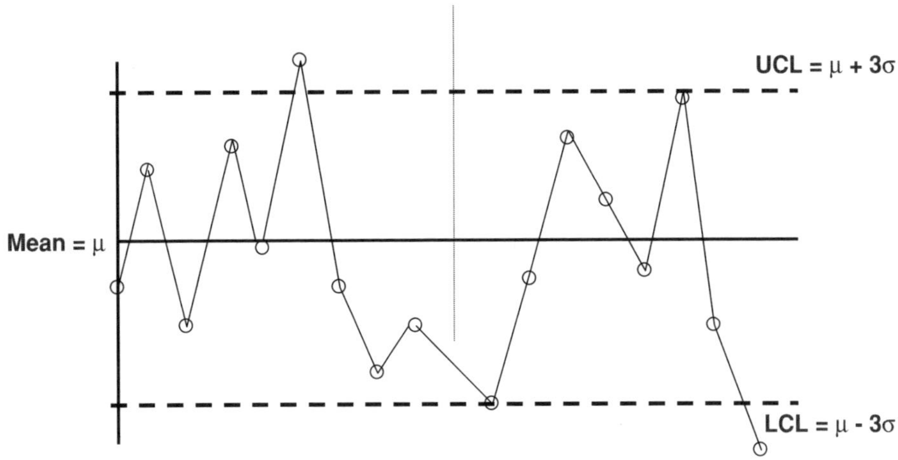

Figure 4.11 Control chart

between them. In Figure 4.11 all data are within the control limits, indicating that the process is stable.

4.3.5.5 Training Requirement

The construction and interpretation of control charts require a good understanding of statistical process control. The concepts of variable data vs attribute data, common causes vs special causes and control limits are essential to the effective application of control charts. Hence, a few hours of classroom training is recommended before the members start the application of this tool. Care should be taken not to confuse a control chart with a run chart.

4.4 TOOLS FOR 'ANALYSE'

The tools for Analyse as described in this section are:

- regression analysis;
- force field analysis;
- SWOT analysis;
- PESTLE analysis;
- five whys;
- inter-relationship diagram.

4.4.1 A1: Regression Analysis

4.4.1.1 Definition

Regression analysis is a tool to establish the 'best fit' linear relationship between two variables.

The knowledge provided by the scatter diagram is enhanced with the use of regression.

4.4.1.2 *Application*

The topic of regression analysis is usually studied at school in algebra lessons, where different techniques of 'curve fitting' are considered. Two common approaches are:

- method of intercept and slope;
- method of least squares.

In a practical business environment, the team members normally resort to drawing an approximate straight line by employing their visual judgement. Sometimes they use the 'method of intercept and slope'. Both of these practices are the estimated 'best fit' relationship between two variables. The reliability of such estimates depends on the degree of correlation that exists between the variables.

Regression analysis is used not only to establish the equation of a line but also to provide the basis for the prediction of a variable for a given value of a process parameter. The scatter diagram, on the other hand, does not predict cause and effect relationships. Given a significant co-relation between the two variables, regression analysis is a very useful tool enabling one to extend and predict the relationship between these variables.

4.4.1.3 *Basic Steps*

We have considered the 'method of intercept and slope' for developing the basic steps as follows (courtesy of Moroney, 1973, p. 284):

1. Consider the equation $y = mx + c$, where m is the slope, c is the intercept, and x and y are the two variables.
2. In the equation $y = mx + c$, substitute each of the pairs of values for x and y and then add the resulting equations.
3. Form a second similar set of equations, by multiplying through each of the equations of Step 2 by its coefficient of m. Add this set of equations.
4. Steps 2 and 3 will each have produced an equation in m and c. Solve these simultaneous equations for m and c.
5. Plot the straight line graph of $y = mx + c$ for the calculated values of m and c.

4.4.1.4 *Worked-out Example*

The following example is taken from Moroney (1973, pp. 278–285).

Consider an investigation made into the relationship between two quantities, y and x, with the following values observed:

y	5	8	9	10
x	1	2	3	4

The values are plotted as shown in Figure 4.12. Now we follow the basic steps to calculate m and c in the equation $y = mx + c$.

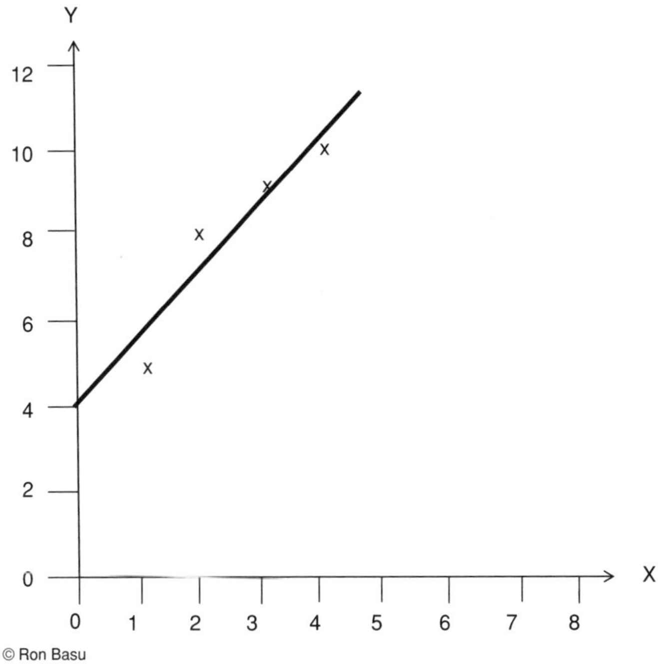

Figure 4.12 Regression analysis

Substituting the observed values of x and y, the resulting equations are:

$$
\begin{aligned}
5 &= m + c \\
8 &= 2m + c \\
9 &= 3m + c \\
10 &= 4m + c \\
32 &= 10m + 4c
\end{aligned}
\tag{4.1}
$$

Multiplying each of the equations by its coefficient of m, the resulting equations are:

$$
\begin{aligned}
5 &= m + c \\
16 &= 4m + 2c \\
27 &= 9m + 3c \\
40 &= 16m + 4c \\
88 &= 30m + 10c
\end{aligned}
\tag{4.2}
$$

We then solve simultaneously Eqs (4.1) and (4.2) for m and c.
We find that

$$m = 1.6$$
$$c = 4$$

Hence

$$y = 1.6x + 4$$

We calculate two pairs of values for x and y to draw a straight line as shown in Figure 4.12.

y	4	8.8
x	0	3

4.4.1.5 Training Requirements

From the above, it will not surprise you to learn that the training of regression analysis is just like brushing up on school algebra! A few classroom exercises for approximately one hour should be adequate for team members to prepare themselves for the practical application of regression analysis.

4.4.2 A2: Force Field Analysis Diagram

4.4.2.1 Definition

The force field analysis diagram, or simply force field diagram, is a model built on the concept by Kurt Lewin (1951). According to this theory, change is characterised as a state of equilibrium between driving forces (e.g. new technology) and opposing or restraining influences (e.g. fear of failure).

In order for any change to occur, the driving forces must exceed the restraining forces, thus shifting the equilibrium.

4.4.2.2 Application

The force field diagram is a useful tool at the early stage of change management leading to improvement. It is often used:

- To investigate the balance of power involved in an issue or obstacle at any level (personnel, project, organisation, network).
- To identify important players or stakeholders – both allies and opponents.
- To identify possible causes and solutions to the problem.

4.4.2.3 Basic Steps

According to Lewin (1951), three key steps are involved in the concept of change management by force field diagram:

- First, an organisation has to unfreeze the driving and restraining forces that hold it in a state of apparent equilibrium.
- Second, an imbalance is introduced to the forces, either by increasing the drivers, reducing the restrainers, or both, to enable the change to take place.
- Third, once the change is complete and stable, the forces are brought back to equilibrium and refrozen.

A force field diagram is constructed by a team and the following basic steps are suggested:

1. Agree on the current problem or issue under investigation and the desired situation.
2. List all forces driving changes towards the desired situation.
3. List all forces resisting changes towards the desired situation.

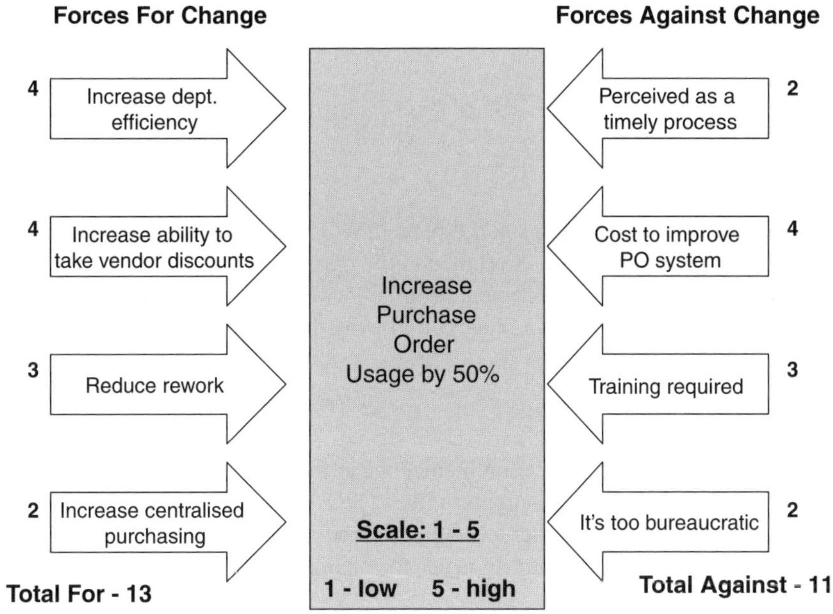

Figure 4.13 Force field analysis diagram

4. Review all forces and validate their importance.
5. Allocate a score to each of the forces using a numeric scale (e.g. 5 = most important and 1 = least important).
6. Chart the influences by listing the driving forces on the left and restraining forces on the right.
7. Decide how to minimise or eliminate the restraining forces and increase the driving forces.
8. Agree on an action plan.

4.4.2.4 Worked-out Example

The issue identified is how to increase the usage of purchase orders in a pharmaceutical company. The driving and restraining forces were identified by the team and also rated on a scale of 1 to 5 (1 = low, 5 = high). This was then represented in a force field diagram as shown in Figure 4.13.

The driving forces show a total score of 13 against an overall score of 11 demonstrated by the restraining forces.

4.4.2.5 Training Requirements

The knowledge and application of a force field diagram are best derived by 'on the job' training during a brainstorming exercise on an actual problem. The facilitator should have experience of change management in applying a number of force field diagrams to direct the team and gain a consensus of identifying forces and allocating scores.

4.4.3 A3: SWOT Analysis

4.4.3.1 Definition

Strengths, weaknesses, opportunities and threats (SWOT) analysis is a tool for analysing an organisation's competitive position in relation to the opposition.

In the context of a quality improvement programme, SWOT analysis refers to a summary of the gaps and positive features of a process following the analytical stage.

4.4.3.2 Application

Based on a SWOT framework, the team members can focus on alternative strategies for improvement. For example, a strengths/opportunities (S/O) approach may be considered to consolidate the strengths and open further leverage from the process. Similarly, a strengths/threats (S/T) policy could be borne in mind in order to maximise the strength of the process and minimise risks.

Thus, a SWOT analysis can help the team to identify a wide range of alternative tactics for the next stage.

4.4.3.3 Basic Steps

1. Create two categories (internal factors and external factors) and then further sub-divide into positive aspects (strengths and opportunities) and negative aspects (weaknesses and threats).
2. Ensure that the internal factors may be viewed as a strength or weakness of a process depending on their impact on the outcome or the process output.
3. Similarly assess the external factors bearing in mind that threats to one process could be viewed as an opportunity for another procedure.
4. Summarise the key features and findings derived from previous analyses in each of the SWOT categories.
5. Develop the improvement strategy for the next stage as pointers from the SWOT analysis.

4.4.3.4 Worked-out Example

The following example is taken from Kotabe and Helsen (2000, p. 277).

Table 4.7 shows the framework of a SWOT analysis.

4.4.3.5 Training Requirements

The training for SWOT analysis should be conducted in stages. The first part takes place in the classroom, where the basic principles can be explained based on a known marketing product.

Table 4.7 SWOT analysis framework

Internal factors / External factors		**Strengths** Brand name, human resources, technology, advertising	**Weaknesses** Price, lack of financial resources, long product development cycle
Opportunities	Growth market, deregulation, stable exchange rate, investment grant	*S/O strategy*: Maximise strengths and maximise opportunities	*W/O strategy*: Minimise weaknesses and maximise opportunities
Threats	New entrants, change in consumer preference, local requirements	*S/T strategy*: Maximise strengths and minimise threats	*W/T strategy*: Minimise weaknesses and minimise threats

The second aspect of the tuition is delivered to the team member when the results following the analysis of process are summarised.

4.4.4 A4: Pestle Analysis

4.4.4.1 Definition

Political, economic, social, technical, legal and environmental (PESTLE) analysis is an analytical tool for assessing the impact of external contexts on a project or a major operation, and also the impact of a project on its external circumstances.

4.4.4.2 Application

Very few significant changes, whether they are caused by a major operation or a project, are unaffected by external surroundings.

Political: A project is affected by the policies of international, national or local government. It is also influenced by company policies and those of stakeholders, managers, employees and trade unions.

Economic: The project is affected by national and international economic issues, inflation, interest rates and exchange rates.

Social: The change is influenced by social issues, the local culture, the lives of employees, communications and language.

Technological: The success of implementation is affected by the technology of the industry and the technical capability of the parent company.

Legal: The project is affected by the legal aspects of planning, registration and working practices.

Environmental: The impact of the change on environmental emission, noise, health and safety is assessed.

4.4.4.3 Basic Steps

PESTLE analysis is carried out in four stages:

1. Develop a good understanding of the deliverables of the operation and the project. At this stage, the relevant policy and guidelines of both the local company and the parent organisation are reviewed.
2. List the relevant factors affecting the various aspects of the project related to PESTLE. It is important that the appropriate expertise of the organisation is drawn into the team for this analysis.
3. Validate the factors in Step 2 with the stakeholders and functional leaders of the company.
4. Review progress and decide on the next steps by asking two questions: How did we do? Where do we go next?
5. For more information on PESTLE analysis, see Turner and Simister (2000, pp. 165–215).

4.4.4.4 Worked-out Example

Consider the policy renewal management process of an insurance company based in Finland. The company implemented an on-line renewal process within Finland and wanted to expand the process in the European Union (EU).

A PESTLE analysis was carried out as shown in Table 4.8.

The PESTLE analysis shows that the expansion of an on-line service, in general, has direct influence upon – and opens up – opportunities in the EU.

4.4.4.5 Training Requirements

The principles of PESTLE analysis are best learned by the process of the group working together during the project life cycle. The programme of Black Belt training normally includes a session on PESTLE analysis and team leaders should receive a broad understanding of this

Table 4.8 PESTLE analysis

Contexts	Key factors	Impact on company (0–10)
Political	Within EU, countries are moving towards a more common political structure.	6
Economic	Slowing economy of Finland; GDP forecast to grow by 3.9% in 2003. Dynamic changes in client business environment in Finland and Europe.	8
Social	Difference in buying habits in Finland vs EU.	7
Technological	Accelerating pace of change in ICT in Finland. On-line opportunity in EU with little extra cost. New cyber-related risk in client business.	9
Legal	New constraints or requirements initiated by regulatory bodies, e.g. Insurance Supervisory Authority in EU.	9
Environmental	No significant impact.	1

tool. The analysis is of a strategic nature and thus may not involve all members of the project team.

4.4.5 A5: The Five Whys

4.4.5.1 Definition

The five whys is a systematic technique of asking five questions successively. The aim is to probe the causes of a problem and thus hopefully get to the heart of the issue.

4.4.5.2 Application

The five whys is a technique that is widely used to analyse problems in both manufacturing and service operations. It is a variation on the classic work study approach of 'critical examination' involving six questions: why, what, where, when who and how?

The objective is to eliminate the root cause rather than merely 'patch up' the effects.

4.4.5.3 Basic Steps

1. Select the problem for analysis.
2. Ask five 'close' questions, one after another, starting with why.
3. Do not defend the answer or point the finger of blame at others.
4. Determine the root cause of the problem.

4.4.5.4 Worked-out Example

The following example is taken from Stamatis (1999, p. 183).

Consider a problem: 'Deliveries are not completed by 4pm'.
Question 1: Why does it happen?
Answer: The routing of trucks is not optimised.
Question 2: Why is it not optimised?
Answer: Goods are loaded based on their size rather than the location of the delivery.
Question 3: Why are they loaded by size?
Answer: The computer defines the dispatch based upon the principle of 'large items first'.
Question 4: Why are large items given preference?
Answer: Large items are delivered first.
Question 5: But why?!
Answer: Current prioritisation policy puts large items first on the delivery schedule.

4.4.5.5 Training Requirements

As can be seen from the above question and answer model, the principle of this analytical tool is very straightforward. Thus, the application of the five whys does not require any rigorous classroom training. The members of a problem-solving team can easily understand and apply this simple tool after just one such group exercise.

4.4.5.6 *Final Thoughts*

The five whys is an uncomplicated but very effective tool that can be used to identify the root causes of a problem. We recommend that, taking advantage of the fact that it is such a quick and unfussy approach, it can be utilised on a far wider basis than at present.

4.4.6 A6: Inter-relationship Diagram

4.4.6.1 *Definition*

An inter-relationship diagram (ID) is an analytical tool to identify, systematically analyse and classify the cause and effect relationships among all critical issues of a process. The key drivers or outcomes are identified, leading to an effective solution.

4.4.6.2 *Application*

An ID is often applied to enable the further examination of causes and effects after these are recorded in a fishbone diagram. The ID encourages team members to think in multiple directions rather than merely in a linear sense.

This simple tool enables the team to set priorities to root causes even when credible data does not exist.

4.4.6.3 *Basic Steps*

1. Assemble the team and agree on the issue or problem for investigation.
2. Lay out all the ideas or issues that have been brought from other tools (such as a cause and effect diagram) or brainstormed.
3. Look for the cause and effect relationships between all issues and assign the 'relationship strength' as:
   3 = significant;
   2 = medium;
   1 = weak.
4. Draw the final ID in a matrix format and insert the 'relationship strength' given by members.
5. Total the relationship strength in each row to identify the strongest effect of an issue on the greatest number of issues.

4.4.6.4 *Worked-out Example*

The following example is taken from Bassard and Ritter (1994, p. 81).

Consider five key issues to improve customer service:

- logistics support;
- customer satisfaction;
- education and training;
- personal incentives;
- leadership.

The ID is plotted in a matrix (see Figure 4.14) with appropriate 'relationship strengths'.

	Logistics Support	Customer Satisfaction	Education and Training	Personal Incentives	Leadership	Total
Logistics Support		○	□	△	□	8
Customer Satisfaction	○		□	○	□	10
Education and Training	□	□		□	○	9
Personal Incentives	△	○	□		○	9
Leadership	□	□	○	○		10

Relationship Strength: Key

○	Significant	3
□	Medium	2
△	Weak	1

Figure 4.14 Inter-relationship diagram

From the above analysis in the 'Total' column, it is evident that customer satisfaction and leadership are the two most critical issues.

4.4.6.5 Training Requirements

The team can be adept in the application of ID after less than one hour's training in a classroom or a practical environment. The principles are simple, but it is important that a consensus is reached in attributing the relationship strength numbers.

4.5 TOOLS FOR 'IMPROVE'

The tools for Improve as described in this section are:

- single minute exchange of dies (SMED);
- five S;
- mistake-proofing;

- brainstorming;
- overall equipment effectiveness (OEE).

4.5.1 I1: SMED

4.5.1.1 Definition

SMED is the name of the approach used for reducing output and quality losses due to change-overs and set-ups.

'Single minute' means that necessary set-up time is counted on a single digit.

4.5.1.2 Application

This method has been developed in Japan by Shigeo Shingo (1985) and has proven its effectiveness in many manufacturing operations by reducing the change-over times of packaging machines from hours to minutes.

The primary application area of SMED is the reduction of set-up times in production lines. This process enables operators to analyse and find out themselves why the change-overs take so long and how this time can be reduced. In many cases, change-over and set-up times can be condensed to less than 10 minutes, so that the duration of the change-over can be expressed with one single digit. It is therefore called 'single minute exchange of dies'.

SMED is considered as an essential tool in Lean Manufacturing and it is instrumental in the reduction of non-value-added activities in process times. Change-over deficit is one of the six big losses that have been defined within the total productive maintenance (TPM). It is important to note that SMED is directly linked with the analytical process of OEE.

With due respect to the success of the SMED method, it is fair to point out that the basic principles are fundamentally the application of classical industrial engineering or work study.

4.5.1.3 Basic Steps

1. Study and measure the operations of the production line to discriminate:
   - internal set-up (IS), the operation that must be done, which machine is stopped;
   - external set-up (ES), the operation that possibly can be done while the machine is still running.
2. Suppress non-value-added operations and convert IS operating into ES. The data from OEE and the preparations of pre-requisites (e.g. tools, change-over parts, pre-assemblies, pre-heating, mobile storage, etc.) are reviewed to achieve results. Some internal set-ups are converted to external set-ups.
3. The next stage is to simplify the design of the machine, especially fillings and tightening mechanisms. Some examples of the design simplification are U-shaped washers, quarter-turn screws, and cam and lever tights.

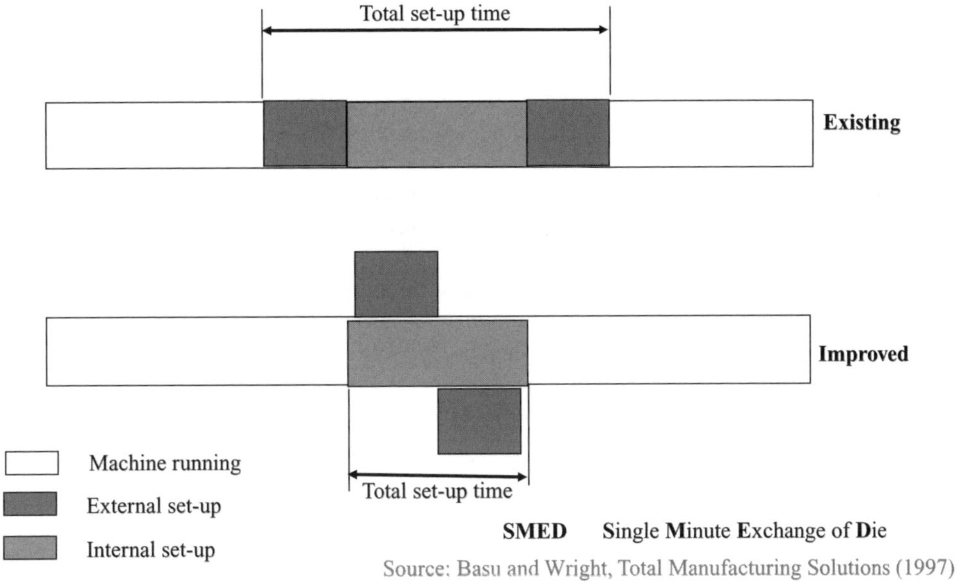

Figure 4.15 SMED: set-up time reduction

4. Balance the work content of the line and ensure team work. For example, in one automatic insertion machine, one operator sets up on the machine front while the other operator feeds components on the other side.
5. Minimise trials and controls. Use of mistake-proofing – or paka-yoke – enables the standard way to be carried out each time.

4.5.1.4 Worked-out Example

The following example is taken from Basu and Wright (1998, p. 97).

Consider the set-up time reduction of a packing machine.

The internal and external set-up times have been measured. As shown in Figure 4.15, the total set-up duration is reduced by overlapping external set-up times on the internal set-up phase.

4.5.1.5 Training Requirements

The training of SMED is likely to be more effective on team members with a good understanding of work study or industrial engineering. The basic principles should be explained to members, with hands-on exercises being carried out in a room. This should be supplemented by a production line study in a factory environment. It is also important that the basics of OEE are covered during a SMED training session.

4.5.2 I2: Five S

4.5.2.1 Definition

Five S is a tool for improving the housekeeping of an operation, developed in Japan, where the five Ss represent five Japanese words all beginning with 'S':

- Seiri (organisation): separate what is essential from what is not.
- Seiton (neatness): sort and arrange the required items in an orderly manner and in a clearly marked space.
- Seiso (cleaning): keep the workstation and the surrounding area clean and tidy.
- Seiketson (standardisation): clean the equipment according to laid down standards.
- Shitsuke (discipline): follow the established procedure.

In order to retain the name 'five S', a number of English-language versions have evolved. These include:

- Seiri: sort;
- Seitor: set in order/stabilise;
- Seiso: shine;
- Seiketsu: standardise;
- Shitsuki: sustain.

4.5.2.2 Application

The five S method is a structured sequential programme to improve workplace organisation and standardisation. Five S improves the safety, efficiency and orderliness of the process and establishes a sense of ownership within the team.

Five S is used in organisations engaged in Lean Sigma, just-in-time (JIT), total productive maintenance (TPM) and total quality management (TQM). This principle is widely applicable not just for the shop floor, but for the office too. As an additional bonus, there are benefits to be found in environmental and safety factors due to the resulting reduction in disorder and confusion. Quality is improved by better organisation and productivity is increased due to the decline in the amount of time spent in searching for the right tool or material at the workstation.

Consider the basic principle of a parent tidying a small child's room which is overflowing with clutter and where parts of some playthings are lost. By sorting through the various types of toys, grouping together pieces of missing jigsaw, Lego™ and so forth to reduce the muddle, eventually no toy or game will be missing a vital part of the set. The end product should be a neater, warmer, brighter and more civilised play environment. The parent hopes that this will encourage the child to utilise all toys and equipment more productively because all relevant pieces are together, space is enhanced and mess is reduced. Till the next time!

It is useful to note that the quality gurus of Japan like numbered lists, e.g. the seven mudas, the five whys, the five Ss. However, the exact number of Ss is less important than observing the simple doctrine of achieving the elimination of waste.

As the five S programme focuses on attaining visual order and visual control, it is thus a key component of visual factory management.

4.5.2.3 *Basic Steps*

1. Sort: the initial step in the five S programme is to eliminate excess materials and equipment lying around the workplace. These non-essential items are clearly identified by 'red-tagging'.
2. Set in order: the second step is to organise, arrange and identify useful items in a work area to ensure their effective retrieval. The storage area, cabinets and shelves are all labelled properly. The objective of this step is, as the old mantra says, 'a place for everything and everything in its place'.
3. Shine: this third action point is sometimes known as 'sweep' or 'scrub'. It includes down-to-basics activities such as painting equipment after cleaning, painting walls and floors in bright colours, and carrying out a regular housekeeping programme.
4. Standardise: the fourth point encourages workers to simplify and standardise the process to ensure that the first three steps continue to be effective. Some of the related activities include establishing cleaning procedures, colour-coding containers, assigning responsibilities and using posters.
5. Sustain: the fifth step is to make five S a way of life. Spreading the message and enhancing the practice naturally involves people and cultural issues. The key activities leading to the success of five S include:
   - Recognise and reward the effort of members.
   - Top management awareness and support.
   - Publicise the benefits.
6. The final step is to continue training and maintaining the standards of five S.

4.5.2.4 *Worked-out Examples*

As five S is primarily a visual process, a good example of promoting its message would be to display pictures of a workplace with photographs showing both 'before' and 'after' depictions of the implementation of five S.

The following example is taken from Skinner (2001) to illustrate the benefits of a five S programme.

Northtrop Grumman Inc. in the USA first deployed five S on a part delivery process. The work area assembled a variety of components into a single product.

Before five S, the area was not well organised, and the process was inefficient. With five S implementations, the area saw a huge 93% reduction in the space employees travel to complete tasks as well as a 42% reduction in the overall floor space.

The system has become a one-piece flow operation between assembly and mechanics, enabling everyone involved to know what the station has and what it needs.

4.5.2.5 *Training Requirements*

Five S is a conceptually simple process, but it requires both initial and follow-up training to inculcate the methodology to all employees. The classroom training sessions should be followed by, as far as practicable, a visit to a site where visual changes due to five S could be observed. A second best option is to show the members photography or videos illustrating the 'before and after' status of the workplace involved in a five S programme.

4.5.3 I3: Mistake-proofing

4.5.3.1 Definition

Mistake-proofing is an improvement tool to prevent errors being converted into defects. It comprises two main activities: preventing the occurrence of a defect and detecting the defect itself.

Mistake-proofing is also known as poka-yoke. The concept was developed by Shigeo Shingo and the term 'poka-yoke' comes from the Japanese words 'poka' (inadvertent mistake) and 'yoke' (prevent).

4.5.3.2 Application

Mistake-proofing is applied in fundamental areas. Although poka-yoke was devised as a component of Shingo's 'zero quality control' for Toyota production lines, it is very easy to understand and grounded in basic common sense.

The process of mistake-proofing is simply paying careful attention to every activity in the process and then placing appropriate checks at each step of the process. Mistake-proofing emphasises the detection and correction of mistakes at the Design stage before they have a chance to become defects. This is then followed by checking. It is achieved by 100% inspection while the work is in progress by the operator and not by the quality inspectors. This inspection is an integral part of the work process.

There are an abundance of examples of simple devices related to mistake-proofing in our everyday surroundings, including limit switches, colour-coding of cables, error-detection alarms, a level crossing gate and many more.

4.5.3.3 Basic Steps

1. Perform Shingo's 'source inspection' at the design stage. In other words, identify possible errors that might occur in spite of preventive actions. For example, there may be some limit switches that provide some degree of regulatory control to stop the machine automatically.
2. Ensure 100% inspection by the operator to detect that an error is either taking place or is imminent.
3. Provide immediate feedback for corrective action. There are three basic actions in order of preference:
   - Control – an action that self-corrects the error, e.g. spell checker.
   - Shutdown – a device that shuts down the process when an error occurs, e.g. a limit switch.
   - Warning – alerts the operator that some error is imminent, e.g. alarm.

4.5.3.4 Worked-out Example

Consider the situation leading to the development of a level crossing. This is a place where cars and trains are crossing paths and the chances of accidents are very high.

The possible errors that might occur would relate to car drivers, who might be preoccupied or distracted while driving (source inspection).

Both the level crossing operator and the car driver should ensure safety features while the work is in progress (judgement inspection).

In order to prevent drivers from making mistakes when a train is approaching, traffic lights were installed to alert the driver to stop (warning).

The lights might not be completely effective, so a gate was installed when a train was coming (shutdown or regulatory function).

The operation of the gate was controlled automatically as the train was approaching (control).

With the above mistake-proofing devices in place, an accident can only occur if either the control and regulatory measures are malfunctioning or the driver wilfully drives around the gate.

4.5.3.5 Training Requirements

There is no 'rocket science' involved in mistake-proofing and it may be perceived in a dismissive fashion: 'that's only common sense'. However, it is critical that there should be some basic training in the principles and applications of mistake-proofing. Furthermore, employees need to be empowered to make improvements in the process by using mistake-proofing. A half-day workshop should be sufficient to meet these training requirements.

4.5.4 I4: Brainstorming

4.5.4.1 Definition

Brainstorming is an improvement tool for a team to generate, creatively and efficiently, a high volume of ideas on any topic by encouraging free thinking.

There are a few variations on the brainstorming process, of which two methods are more frequently used. First is the structured approach (known as the 'round robin') where each member is asked to put forward an idea. The other technique is unstructured and is known as 'free-wheeling', in which ideas are produced and expressed by anyone at any time.

4.5.4.2 Application

Brainstorming is employed when the solution to a problem cannot be found by quantitative or logical tools. It works best by stimulating the synergy of a group. One member's thoughts trigger the idea of another participant, and so on. It is often used as a first step to open up ideas and explore options, and these are then followed up by appropriate quality management tools and techniques.

It has the advantage of getting every member involved, avoiding a possible scenario where just a few people dominate the whole group.

There are some simple ground rules or codes of conduct to observe:

- Agree to a time limit with the group.
- Accept all ideas as given and do not interpret or abbreviate.
- Do not evaluate ideas during the brainstorming process.
- Encourage quantity rather than quality of ideas.
- Discourage the role of an expert.

- Keep ideas expressed in just a few words.
- Emphasise causes and symptoms as opposed to solutions.
- Write clearly and ensure the ideas are visible to everyone.
- Have fun!

4.5.4.3 Basic Steps

1. Clearly state the focused problem selected for the brainstorming session.
2. Form a group and choose a facilitator, agree on a time limit and remind members of the ground rules.
3. Decide whether a structured approach or a free-wheeling basis will be used. For a larger group, a structured process will allow everyone to get a turn and subsequently this could be switched to the free-wheeling method.
4. Write clearly on a flipchart or a board any ideas as they are suggested. The facilitator will motivate and encourage participants by prompting them, 'What else?'
5. Review the clarity of the written list of ideas, allow them to settle and discard any duplication.
6. Apply filters to reduce the list. Typical filters could include cost, quality, time and risk.
7. Ensure that everyone concurs with the shortlist of ideas.

4.5.4.4 Worked-out Example

Consider the following focused statement for brainstorming:

'What are the key selection criteria of a family holiday?'

The five members of a family generated 26 ideas or issues. These were then filtered by a budgeted cost of £4000 for the whole family and the following key criteria were derived:

- Two weeks in August.
- Seaside resort.
- Indoor and outdoor recreational facilities.
- Not near a nightclub.
- Rich local culture.
- Opportunities for sightseeing.

4.5.4.5 Training Requirements

The application of brainstorming does not require any formal training in a classroom. A facilitator with some previous experience in the process can conduct a successful brainstorming session after briefing the team with the ground rules.

4.5.5 I5: OEE

4.5.5.1 Definition

The overall equipment effectiveness (OEE) is an index measuring the delivered performance of a plant or equipment based on good output.

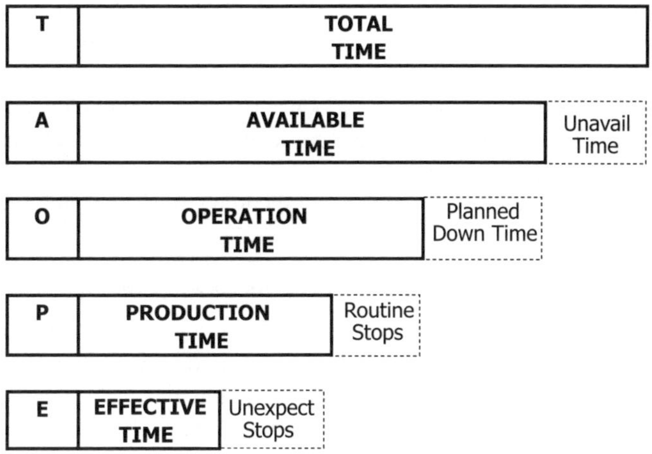

Figure 4.16 Equipment time analysis

The method of monitoring OEE is devised in such a way that it would highlight the losses and deficiencies incurred during the operation of the plant and identify the opportunities for improvement.

There are many ways to calculate OEE (see Shirose, 1992; Hartman, 1991). In this section we describe the methodology of OEE that was developed and applied by the author at both Unilever[1] and GlaxoWellcome.[2]

OEE is defined by the following formula:

$$\text{OEE (\%)} = \frac{\text{Actual good output} \times 100}{\text{Specified output}}$$

where:

$$\text{Specified output} = \text{Specified speed} \times \text{Operation time.}$$

4.5.5.2 Application

The application of OEE has been extensive, especially when driven by the TPM (total productive maintenance) programmes, to critical plant and equipment. It can be applied to a single item of equipment, a packing line, a production plant or processes. In order to appreciate the usefulness of OEE it is important to understand equipment time analysis as shown in Figure 4.16 and described below.

'Total time' defines the maximum time within a reporting period, such as 52 weeks a year, 24 hours a day, 8760 hours a year.

'Available time' is the time during which the machine or equipment could be operated within the limits of national or local statutes, regulation or convention.

'Operation time' is the period during which the machine or equipment is planned to run for production purposes. The operational time is normally the shift hours.

[1] At Unilever plc, the methodology was known as PAMCO (plant and machine control).

[2] At GlaxoWellcome it was called CAPRO (capacity analysis of production).

'Production time' is the maximum phase during which the machine or equipment could be expected to be operated productively after adjusting the operation time for routine stoppages such as change-over and meal breaks.

'Effective time' is the duration needed to produce a 'good output delivered' if the machine or equipment is working at its specified speed for a defined period. It includes no allowances for interruptions or any other time losses.

It is important to note that effective time is not recorded, it is calculated from the specified speed as:

$$\text{Effective time} = \text{Good output}/\text{Specified speed}$$

where specified speed is the optimum speed of a machine or equipment for a particular product without any allowances for loss of efficiency. It is expressed as quantity per unit (such as tons per hour, bottles per minute, cases per hour or litres per minute).

In addition to OEE, two other indices are commonly used as shown below:

$$\text{Production efficiency } (\%) = \frac{\text{Effective time } (E) \times 100}{\text{Production time } (P)}$$

$$\text{Operational utilisation } (\%) = \frac{\text{Operation time } (O) \times 100}{\text{Total time } (T)}$$

A properly designed and administered OEE scheme offers a broad range of benefits and a comprehensive manufacturing performance system. Some of its key benefits are:

- It provides information for shortening lead time and change-over time and a foundation for SMED.
- It provides essential and reliable data for capacity planning and scheduling.
- It identifies the 'six big losses' of TPM, leading to a sustainable improvement in plant reliability.
- It provides information for improving asset utilisation and thus reduced capital and depreciation costs in the longer term.

4.5.5.3 Basic Steps

1. Select the machines, equipment or production line where the OEE scheme could be applied. The selection criteria will depend on the criticality of the equipment in the context of the business. It is useful to start with a single production line as a trial or pilot.
2. Establish the specified speed of the production line governed by the control or bottleneck operation. Figure 4.17 shows an example of a soap packaging line, where the specified speed is 150 tablets per minute, i.e. this constitutes the speed of the wrapper (which is the slowest piece of equipment).

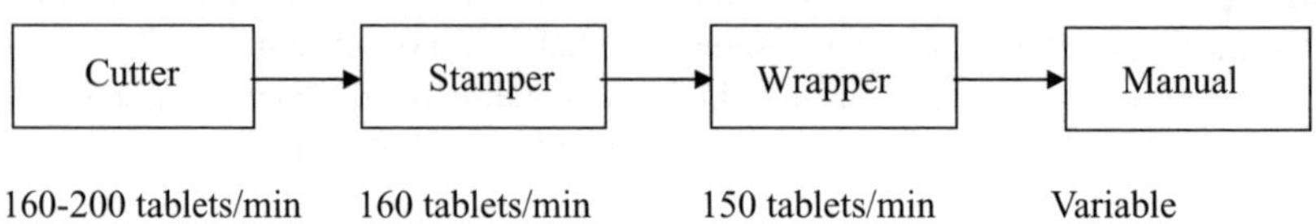

Source: Basu and Wright (2003).

Figure 4.17 Soap production line

3. Set up a data recording system so that the output data and various stoppages and losses can be recorded.
4. Compile the data every day and validate the results. At this stage, detailed calculations are not necessary.
5. Monitor the results, comprising OEE and key indices, major losses as a percentage of the operation time and the trends of indices. The reporting is normally on a weekly basis for the department and on a monthly basis for senior management.
6. Use the results for continuous improvement, planning and strategic changes.

4.5.5.4 Worked-out Example

Consider the production data of a toilet soap packing line (Table 4.9), where the control station governing the specified speed is an ACMA 711 wrapping machine.

Given that each case contains 144 tablets:

$$\begin{aligned}
\text{Good output} &= 4232 \times 144 = 609{,}408 \text{ tablets}\\
\text{Effective time} &= \frac{\text{Good output}}{\text{Specified speed}} = \frac{609{,}408}{150 \times 60} = 67.71 \text{ hours}\\
\text{Production time} &= \text{Operation time} - \text{Routine stoppages}\\
&= 128 - 11.5 = 116.5 \text{ hours}\\
\text{Total time} &= 7 \times 24 = 168 \text{ hours}\\
\text{OEE} &= \frac{\text{Effective time}}{\text{Production time}} = \frac{67.71}{128} = 0.53 = 53\%\\
\text{Production efficiency} &= \frac{\text{Effective time}}{\text{Production time}} = \frac{67.71}{116.5} = 58\%\\
\text{Operation utilisation} &= \frac{\text{Operation time}}{\text{Total time}} = \frac{128}{168} = 76\%
\end{aligned}$$

It is important to note that the effective time was calculated and not derived from the recorded stoppages. There will be an amount of unrecorded time (also known as time adjustment) given by, in the example:

$$\begin{aligned}
\text{Unrecorded time} &= (\text{Production time} - \text{Unexpected stoppages}) - \text{Effective time}\\
&= (116.5 - 27.25) - 67.71\\
&= 21.54 \text{ hours}
\end{aligned}$$

Table 4.9 Soap packing line production data

Week number:	31
Operation time:	128 hours
Specified speed:	150 tablets per minute
Good output:	4232 cases
Routine stoppages:	11 hours 30 minutes
Unexpected stoppages:	27 hours 15 minutes

4.5.5.5 Training Requirements

The success of an OEE scheme depends heavily on the rigour of continuous training. It is important that each operator, supervisor and manager of a production department receives a half-day training programme covering the definitions, purpose and application of the OEE scheme. The training is continuous because of the turnover of staff. Senior management should also benefit from a one-hour awareness session.

4.6 TOOLS FOR 'CONTROL'

The tools for Control as described in this section are:

- Gantt chart;
- activity network diagram;
- radar chart;
- PDCA cycle.

4.6.1 C1: Gantt Chart

4.6.1.1 Definition

A Gantt chart is a simple tool which represents time as a bar or a line on a chart. The start and finish periods for activities are displayed by the length of the bar and often the actual progress of the task is also indicated.

A Gantt chart is also known as a bar chart.

4.6.1.2 Application

The most common form of scheduling is the application of Gantt charts. The merits of Gantt charts are that they are simple to use and provide a clear visual representation of both the scheduled and actual progress of activities. The current time is also indicated on the graph.

In addition, Gantt charts are used to review alternative schedules by using movable pieces of paper or plastic channels. The graphs can be drawn easily by standard software tools such as PowerPoint or Excel. However, a Gantt chart is not an optimising tool and therefore does not determine the 'critical path' of a project.

4.6.1.3 Basic Steps

1. Identify the key activities or the tasks related to the project and describe each action using selective keywords.
2. Prepare a scheduling board, and depending on the duration of the project, draw vertical lines to divide the board into monthly, weekly or daily intervals.
3. Arrange the activities in a sequence of estimated start dates and post them on the extreme left-hand column of the board.
4. Estimate the start and finish dates of each activity and draw horizontal bars or lines along the time scale to reflect the start and duration of each of the activities.
5. On completion of each activity, show the actual start and duration of the activity by using a bar of a different colour.

6. Include a 'time now' marker on the chart; review and maintain the chart until the end of the project.

Activity	Week 1	Week 2	Week 3	Week 4	Week 5	Week 6	Week 7	Week 8	Week 9
A	2	2							
B			2	2					
C			6	6					
D					3	3			
E					1	1	1	1	
F									2
Total resources	2	2	8	8	4	4	1	1	2

Source: Basu and Wright (2003).

Figure 4.18 A Gantt chart

4.6.1.4 Worked-out Example

Figure 4.18 gives an example of a Gantt chart, showing the planned and completed activities of a FIT SIGMA programme.

4.6.1.5 Training Requirements

The application of Gantt charts does not require extensive training. Team members are usually experienced in the use of Gantt charts. A briefing session in front of the scheduling board should be adequate.

4.6.2 C2: Activity Network Diagram

4.6.2.1 Definition

An activity network diagram is a control tool to determine and monitor the most efficient path, known as the critical path, and a realistic schedule for the completion of a project. The diagram is represented graphically showing a brief description of all tasks, their sequence, their expected completion time and the jobs that can be carried out simultaneously.

An activity network diagram with some variations is also referred to as PERT (project evaluation and review technique), CPM (critical path method), a precedence diagram and, finally, network analysis.

4.6.2.2 Application

The activity network diagram was used extensively in most projects during the 1960s and 1970s. As larger undertakings became increasingly complex and comprised numerous tasks, its popularity by manual methods started to diminish. However, with the advent of software systems such as Primavera and MS Project, its application at the higher level of the project has increased significantly. It offers a number of benefits:

- The team members can visualise the criticality of major tasks in the overall success of the project.
- It highlights the problems of 'bottlenecks' and unrealistic timetables.
- It provides facilities to review and adjust both the resources and schedules for specific tasks.

4.6.2.3 Basic Steps

There are normally two methods applied for the construction of an activity network diagram: the 'activity on arrow' method and the 'activity on node' approach. It is the former that has been used most widely, and the steps for the application of the arrow technique can be outlined as follows:

1. Assemble the project team with the ownership and knowledge of key tasks.
2. List the key tasks with a brief description for each one.
3. Identify the first task that must be carried out, the tasks that can be done in parallel and the sequential relationship between tasks.
4. Draw arrows for each task which are labelled between numbered nodes and estimate a realistic time for the completion of each of these tasks.
5. Avoid feedback loops in the diagram. Unlike Gantt charts, the length of the arrows does not have any significance.
6. Determine the longest cumulative path as the critical path of the project.
7. Review the activity network diagram and adjust resources and schedules if appropriate.

For more detailed information on the activity network diagram, see Wild (2002, pp. 403–450).

4.6.2.4 Worked-out Examples

Consider a project of writing and submitting the draft manuscript of a technical book such as this one to a publisher. Table 4.10 lists all the activities which constitute such an endeavour, including their dependent relationship and estimated duration.

The activity network diagram is shown in Figure 4.19.

The critical path is A, B, E, F, G, J, K and M with a total project duration of 22 weeks.

4.6.2.5 Training Requirements

The basic principles of an activity network diagram are not difficult to follow and can be covered in a half-day workshop with practical exercises.

A detailed analysis of an activity network diagram can be complex when the variable estimates of duration, 'float' or 'slack', resource levelling and the probability of occurrence

Table 4.10 List of project activities for production of a technical book

Resource	Activity	Description	Predecessor	Duration (weeks)
Author	A	Prepare proposal	—	2
Publisher	B	Approve proposal	A	4
Author	C	Preliminary research	A	2
Author	D	Detailed research	C	10
Author	E	Write Chapters 1–3	B	3
Author	F	Write Chapters 4–6	E	3
Author	G	Write remaining chapters	F	4
Admin	H	Type Chapters 1–3	E	2
Admin	I	Type Chapters 4–6	F, H	2
Admin	J	Type remaining chapters	G, I	3
Author	K	Compile full draft	D, J	2
Author	L	Obtain copyright clearance	B	12
Author	M	Submit manuscript	K, L	1

are considered. We suggest that the advanced applications should be reserved for academic research.

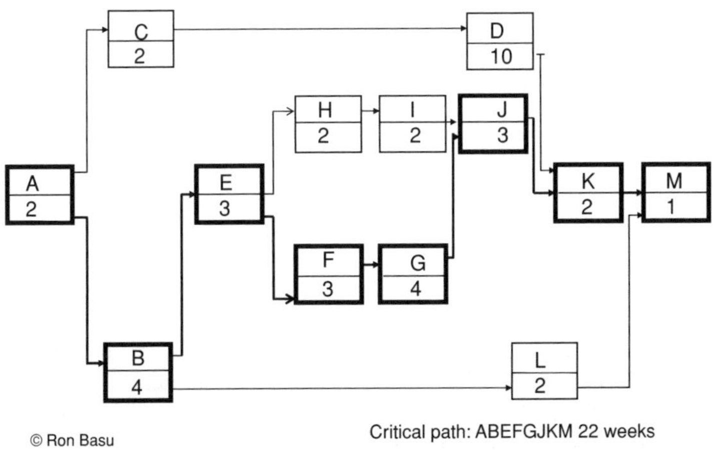

Figure 4.19 Activity network diagram

4.6.3 C3: Radar Chart

4.6.3.1 Definition

A radar chart is a polar graph to show, using just one graphic, the size of the gaps in the performance levels of key performance indicators.

A radar chart is also known as a polar graph and, because of its appearance, as a spider diagram.

4.6.3.2 *Application*

A radar chart is a useful visual tool to display the important metrics of performance at the Control stage of a quality improvement programme. The other benefits of this chart include:

- It highlights the strengths and weaknesses of the total process, programme or organisation.
- It can define full performance in each category.
- It can act as a focal point to capture and review the different perception of all stakeholders of the organisation related to relevant performance metrics.
- Given a range of rating (say on a scale of 1 to 5), it can drive a total or average score of all entities.

However, a limitation of the radar chart is that it tends to provide just a snapshot of the performance levels at any given time.

4.6.3.3 *Basic Steps*

1. Select and define the performance categories. The chart can handle 10–20 categories.
2. Some performance metrics are likely to be easily quantifiable and expressed as a percentage. Other metrics may be qualitative and not represented by numbers.
3. Normalise all performance metrics on a scale of 1 to 5 with appropriate guidelines according to the following grades:

   1 = poor;

   2 = fair;

   3 = good;

   4 = very good;

   5 = excellent.

4. Construct the chart by drawing a wheel with as many spokes as the performance categories and marking each spoke with '0' at the centre and '5' on the rim.
5. Connect the ratings for each performance category and highlight the gaps.
6. Use the results for consolidating strengths and improving weaknesses.

4.6.3.4 *Worked-out Example*

The following example of a radar chart is taken from Basu and Wright (1998, p. 199).

Consider that all the operations of a manufacturing organisation can be expressed in six categories and 20 metrics, as given in Table 4.11.

Figure 4.20 shows a radar chart depicting a snapshot of 20 metrics.

4.6.3.5 *Training Requirements*

The application of a radar chart does not require any significant classroom-based training. However, it does necessitate a good understanding of how to rate all performance categories.

Table 4.11 The operations of a manufacturing organisation

Category (pillars)	Metrics (foundation stones)
Marketing and innovation	3
Environment and safety	3
Supply chain management	3
Manufacturing facilities	5
Procedures	3
People	3

This awareness, and the selection criteria of the metrics, should be explained in a period of about an hour as part of a training programme.

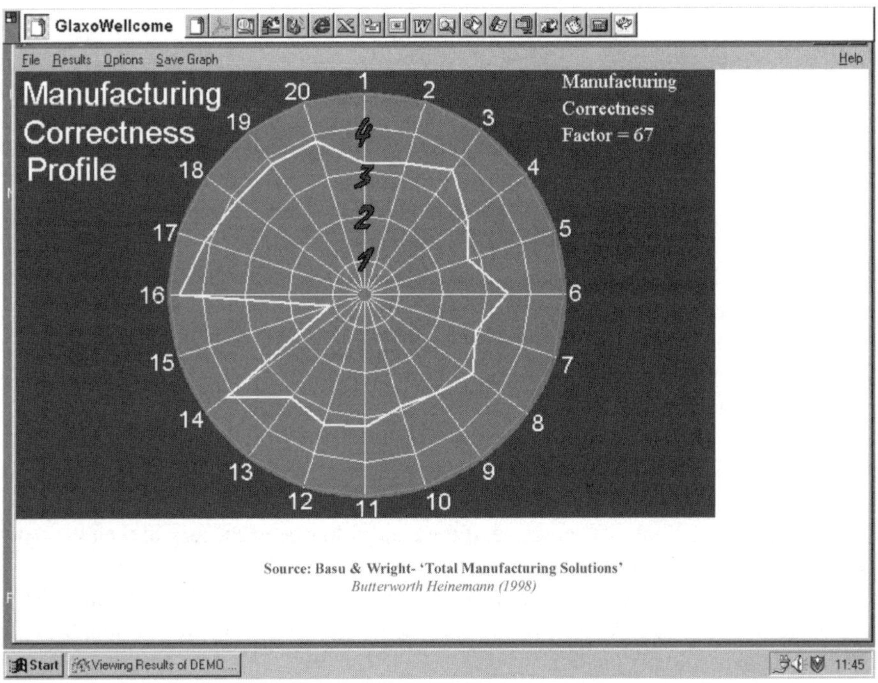

Figure 4.20 A radar chart

4.6.4 C4: The PDCA Cycle

4.6.4.1 Definition

In a central process, the actual results of an action are compared with a target or a set point. The difference between the two is then monitored and corrective measures are adopted if the disparity becomes large. The repeated and uninterrupted nature of continuous improvement follows this usual definition of Control and is represented by the PDCA (plan, do, check, act) cycle.

This is also referred to as the Deming wheel, named after W. E. Deming. Another variation of PDCA is PDSA (plan, do, study, act).

4.6.4.2 Application

The application of the PDCA cycle has been found to be more effective than adopting the 'right first time' approach of concentrating on developing flawless plans (Juran, 1999, p. 41.3). The PDCA cycle means continuously looking for better methods of improvement.

The PDCA cycle is effective in both doing a job and managing a programme. The extent to which the PDCA cycle is applied to the task level depends on the self-control of the operators. Education and training enhance the self-control capacity of workers. At the programme level, the PDCA cycle acts as a process of repeatedly questioning the detailed working of the operations and thereby helps to sustain the improved results.

The PDCA cycle enables two types of corrective action – temporary and permanent. The temporary action is aimed at results by practically tackling and fixing the problem. The permanent corrective action, on the other hand, consists of investigating and eliminating the root causes and thus targets the sustainability of the improved process.

4.6.4.3 Basic Steps

1. P (Plan) stage: the cycle starts with the 'plan' stage, comprising the formulation of a plan of action based on the analysis of the collected data.
2. D (Do) stage: the next step is the 'do' or implementation period. This may involve a mini-PDCA cycle until the issues of implementation are resolved.
3. C (Check) stage: the next step is the 'check' phase where the results after implementation are compared with targets to assess if the expected performance improvement has been achieved.
4. A (Act) stage: at the final 'act' juncture, if the change has been successful then the outcome is consolidated or standardised.
5. If the change has not been successful, however, the lessons are recorded and the cycle starts again. Even if the change is successful, the results are sustained by going through the PDCA cycle over and over again.

4.6.4.4 Worked-out Example

The following example is taken from Juran (1999, pp. 32.8–32.10).

In this illustration, a healthcare organisation in the USA was adhering to the traditional reliance on extensive internal and external inspection to maintain quality standards. This resulted in a medical record system whose size, complexity and format were wasteful and cumbersome.

The aspects of the PDCA cycle (see Figure 4.21) were applied to their internal quality assurance procedures and the medical record procedures were simplified.

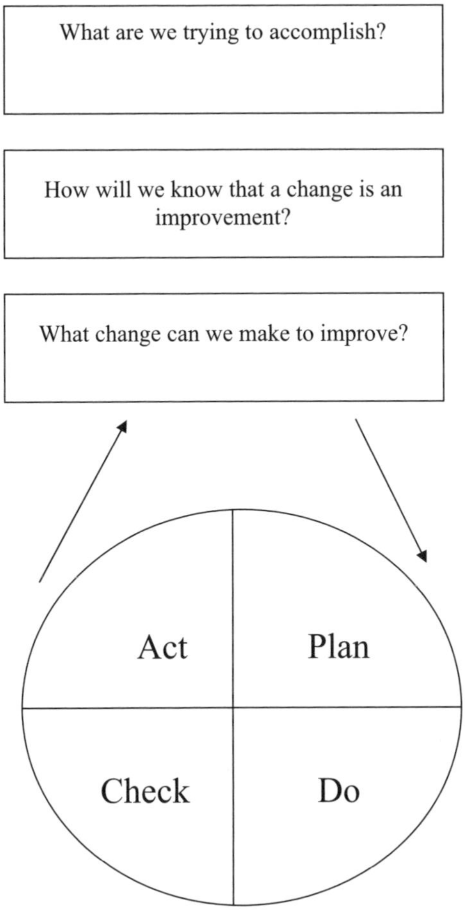

Figure 4.21 PDCA cycle

4.6.4.5 Training Requirements

The importance of training for the PDCA cycle, especially for first-time workers, has been well recognised at the early stage of the quality movement and thus the concept of the 'quality circle' was born. During the late 1950s, about 100,000 transcripts of radio broadcast text for quality circles were sold in Japan. The teaching of the PDCA cycle, although simple in principle, should be inculcated in everyone within the organisation on a continuous basis. This tool then becomes most effective when it becomes a way of life in the organisation.

4.7 SUMMARY

The success of a quality programme is underpinned by the selection and application of appropriate tools and FIT SIGMA is no exception to that. The recommendation is to start with simple tools first, as described in this chapter. Arguably, in this section, only the control chart would require a good understanding of statistical process control. Other tools can be applied without the 'fear' of advanced statistics associated with Six Sigma. It is recognised that more advanced tools and techniques (e.g. DOE, FMEA, QFD, SPC, DFFS, Monte-Carlo simulation and TRIZ) are required to solve more complex problems to near perfection. A separate book, *Implementing Six Sigma and Lean* (Basu, 2009), is recommend to help the reader tackle this area.

5
FIT SIGMA in Large Manufacturing Operations

5.1 INTRODUCTION

It is well accepted that in the 1970s Japanese firms set the pace with their focus on quality and performance reviews. By comparison, western companies (that is, US and European owned and operated businesses) appeared to be less concerned about the importance of quality measures and the impact of standards of performance on bottom line results. TQM initiatives that were taken seemed to concentrate on softer cultural issues rather than 'hard' performance standards. However, with the introduction of Six Sigma tools in the 1980s, Motorola revolutionised the quality movement. Western companies that operated at levels of Two to Three Sigma (with between 45,500 and 2700 defects per million operations) became increasingly interested in improving their performance standards (and their share price). 99% might sound impressive, but it slowly dawned that there is a tremendous difference between 99% and 99.9997%. For example, for every million articles of mail the difference is between 10,000 lost items and three lost items. Or, for two million prescriptions of medicine per annum, 99% = 20,000 wrong prescriptions whereas 99.9997% would in theory equate to seven wrong prescriptions (and in 20 million opportunities the difference is from 200,000 errors to 68). In practice, 99.9997% would mean no wrong prescriptions as the process and culture is conditioned for zero defects, rather than a process or culture that accepts it as inevitable, and acceptable, that mistakes will occur.

The first wave of Six Sigma, following the grand groundwork carried out by Motorola, included Allied Signal, Texas Instruments, Ratheon and Polaroid (to name a few). GE entered the arena in the mid-1990s and in turn was followed by many powerful corporations including SONY, HOB, Dow, Bombadier and GSK. The ability to leverage the experience of successful Six Sigma players proved highly attractive, both as a competitive issue and also to improve profit margins. The following case examples provide insights for organisations that are achieving success in their business performance through the use of Six Sigma programmes.

5.2 FITNESS FOR THE PURPOSE

We talk of Six Sigma level implementations having an accuracy of 99.99966%. Do we need to be so accurate? Isn't it enough to be accurate to, say, 99% of the time? According to the American Society for Quality, being 99% accurate means that:

- As many as 50 babies are dropped in hospitals every day.
- There is no electric power for nearly seven hours every month.
- Every hour, at least 20,000 letters are lost in the mail.
- Over 200,000 errors are made in medical prescriptions on an annual basis.
- Every week, doctors botch up 5000 surgeries.
- Major airports report two incorrect landings every day.

Now, doesn't that additional 0.99966% (Six Sigma) sound worthwhile? Well, the answer to the question is 'that depends on the application'. For example, for a surgeon's operation or aircraft manufacture, the accuracy of Six Sigma is justified, while for the manufacture of bin bags an accuracy of One Sigma is adequate.

Many large organisations' manufacturing precision (e.g. aircraft systems) and life-saving (e.g. pharmaceutical) products should allocate adequate resources to aiming towards achieving Six Sigma accuracy. This approach conforms to the FIT SIGMA tenet of 'fitness for the purpose'.

5.3 SIGMA (Σ) FOR IMPROVEMENT AND INTEGRATION

The focus on a data-driven approach of Six Sigma implies that the process information is available. If this is not the case, it is possible to waste a lot of resources in obtaining process data. The key components of this FIT SIGMA tenet for large manufacturing organisations (e.g. appropriate DMAIC methodology, learning deployment, project plans, etc.) are similar to those of a comprehensive Six Sigma or Lean Sigma programme. However, the FIT SIGMA approach recognises that even for a large organisation it may over-complicate the investigation if advanced techniques are used on simple problems. The following two case studies illustrate this point of appropriate application. Although General Electric and Seagate did not adopt the name FIT SIGMA for their programmes, their applications of Six Sigma principles do conform to the laws of FIT SIGMA.

5.3.1 Case Study One: General Electric

With over 4000 Black Belts and 10,000 Green Belts across its businesses, and Six Sigma savings of $2 billion in 1999 alone, GE is both a comprehensive Six Sigma organisation and can be seen as a benchmark for Six Sigma programmes.

5.3.1.1 The Company

General Electric has been at the top of the list of *Fortune* 500's most admired companies for the last five years, and without doubt their Six Sigma programme has played a key role in their continued success. In 2001, GE's turnover was over $125.8 billion, they employed 310,000 people world-wide and their market value was $401 billion. With earnings growing at 10% per annum, GE also had the enviable record of pleasing Wall Street and financial analysts year after year. GE's products and business categories span a wide spectrum of automotive, construction, healthcare, retail, transport, utilities, telecommunications and finance.

5.3.1.2 Driver for Six Sigma

The CEO of GE, Jack Welch, is reported to have become attracted to the systematic and statistical method of Six Sigma in the mid-1990s. He was ultimately convinced of the power of Six Sigma after a presentation by Allied Signal's former CEO, Larry Bossidy, to a group of GE employees. Bossidy, a former Vice Chairman of GE, had witnessed excellent returns from Allied Signals' experience with Six Sigma.

In 1995, GE retained the Six Sigma Academy, an organisation started by two early pioneers of the process, both ex-Motorola – Michael Harry and Richard Schroeder. It was pointed out that the gap between Three Sigma and Six Sigma was costing GE between $7 and $10 billion annually in scrap, rework, transactional errors and lost productivity. With the full and energetic support of Jack Welch, senior management became fully committed to the Six Sigma programme. 'GE Quality 2000' became the GE mantra for the 1990s and beyond. Jack Welch declared that 'Six Sigma, "GE Quality 2000" will be the biggest, the most personally rewarding and in the end the most profitable undertaking in our history!' While financial benefits and the share price were a driving force in Six Sigma deployment, GE identified four specific reasons for implementing Six Sigma:

1. cost reduction;
2. customer satisfaction improvement;
3. Wall Street recognition;
4. corporate synergies.

5.3.1.3 Improvement Programme

Although Motorola pioneered the Six Sigma programme in the 1980s to improve manufacturing quality and to eliminate waste in production, GE broke the mould of Motorola's original process by applying Six Sigma standards to its service-oriented businesses – GE Capital Services and GE Medical Systems. Note that GE Capital Services accounts for nearly half of GE's total sales.

The structure of a typical GE Six Sigma team is shown in Figure 5.1.

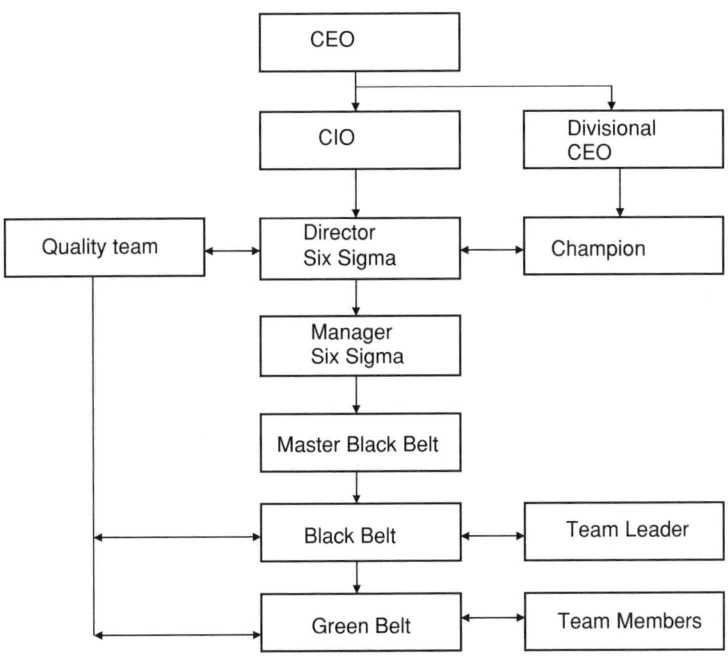

Figure 5.1 The structure of a GE Six Sigma team

The Six Sigma programme was launched in 1995 with 200 separate projects supported by a massive training effort. In the following two years a further 9000 projects were successfully undertaken and the reported savings were $600 million. The training investment for the first five years of the programme was close to $1 billion. GE also instituted a personnel recruitment plan to augment the cadre of dedicated full-time Six Sigma staff.

The GE programme revolved around the following few key concepts, all focused on the customer and internal processes.

1. Critical to quality: determination and development of attributes most important to the *customer*.
2. Defect: identification of failure to meet *customer* wants.
3. Process capability: what the *process* can deliver.
4. Variation: what the *customer* sees and feels as against what the customer wants.
5. Stable operation: ensuring consistent and predicable *processes* to improve what the customer sees and feels.
6. Design for Six Sigma: designing to meet *customer* needs and *process* capability.

5.3.1.4 Model for Rollout

There does not appear to be one universal model for the rollout of Six Sigma amongst the companies that have implemented a Six Sigma programme. However, the Six Sigma Academy has advised that there is a general paradigm which is effective and has been adopted/developed by GE. This broad-spectrum model is shown in Table 5.1.

5.3.1.5 Key Benefits Achieved

Figure 5.2 shows the direct financial benefits achieved by GE over a four-year period. This provides evidence of the very real benefits that are achievable from a Six Sigma programme. With FIT SIGMA the next stage is to sustain and grow this payback.

At the second level of benefits, where the impact on savings is not direct, the achievements, stated as an average per year, include:

20% margin improvement;
12–18% capacity increase;
12% reduction in headcount;
10–30% capital expenditure reduction.

Table 5.1 GE training model

Phase one	Business units select champions and Master Black Belts. The Six Sigma Academy recommends one Champion per business group and one Master Black Belt for every 30 Black Belts.
Phase two	Champions and Master Black Belts undergo training. The over-riding deployment plan is developed.
Phase three	Champions and Master Black Belts, with the assistance of Black Belts, begin identifying potential projects.
Phase four	Master Black Belts receive additional training, focusing on how to train other staff.
Phase five	Black Belts undergo training and the first projects are officially launched.
Phase six	Black Belts begin training Green Belts.

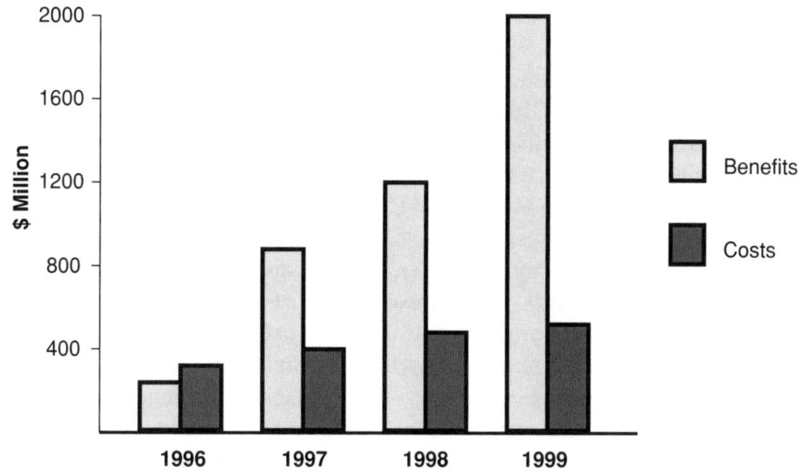

Figure 5.2 Six Sigma payoff at GE

Some specific examples from business units are:

GE Medical Systems – in the introductory year there were 200 successful projects.
GE Capital – invested $6 million over four years to train just on 5% of the work force who were involved full time on quality projects, 28,000 of which were successfully completed.
GE Aircraft Engines – the time taken to overhaul engines reduced by an average of 65 days.
GE Plastics – in just one project, a European polycarbonate unit increased capacity by 30% in eight months.

Market consultants and analysts have reacted favourably to GE's achievements with Six Sigma. Merrill Lynch is quoted as saying, 'Six Sigma balance sheet discipline plus service and global growth are helping fuel (GE's) 13% earning per share gains'.

On 8 May 2002, GE advised that they would deliver record earnings in 2002 of more than $16.5 billion, and were comfortably forecasting double-digit earnings growth for 2003.

5.3.1.6 Lessons Learnt

At one level, the lofty aim of emulating GE may be considered as being beyond the reach of many companies. It is cash-rich, and its business generates over $10 billion per month ($125.8 billion sales for 2001). It makes real things like turbines and refrigerators, and people buy their products with real money. However, on closer examination there are some strong learning points from the GE Six Sigma programme that can benefit ANY company embarking on a quality programme. These learning points include:

1. Leadership support. There is absolutely no doubt from published data that the chief architect of success was Jack Welch. For any organisation wanting to change a culture such as that required for Six Sigma and FIT SIGMA, strong unstinting leadership from the

top is essential. Likewise, all senior management have to be engaged and believe in the philosophy.

2. Definition of Six Sigma objectives. The aim is to be 'World Class'. So what does this actually mean? Firstly, World Class companies such as GE recognise that quality initiatives are synonymous with profit enhancement and share price. In addition, World Class means internal efficiency, best practice and a focus of customer satisfaction. The lesson is that anything less than an ambition to be World Class simply won't do.
3. Development of initial processes and tools. At GE each problem was defined through measurement and analysis along a five-step DMAIC approach, and the use of the seven quality 'tools' of control charts, defect measurement, Pareto analysis, process mapping, root cause analysis, statistical process control and the decision tree diagram. The lesson is that a structured approach has to be followed for Six Sigma process management.
4. Alignment of Six Sigma with career paths. At GE, Black Belt status became essential for staff on the fast track for advancement. Black Belts were rewarded with share options (in most companies such share options are reserved for senior management). The lesson from this is that recognition has to be given to motivate and retain valuable talent.
5. Six Sigma and service industries. The piloting of Six Sigma in GE Medical Systems and GE Capital Services incontrovertibly proved that Six Sigma is not just for manufacturing; the process is equally applicable to all operations including services. GE has opened the gate for service operations. In western economies 80% of gross domestic product is from the service sector.

Source: Best Practices LLC (2002).

5.3.2 Case Study Two: Seagate Technology International

5.3.2.1 Company Overview

Seagate's position as the world's largest manufacturer of disc drives, magnetic discs and read/write heads, and as a leader in storage area network (SAN) solutions, puts it at the heart of today's 'information-centric' world. Throughout its history, since its foundation in 1979, Seagate has successfully relied on a strategy of vertical integration: designing, developing and producing the key, enabling technologies that go into its storage products, rather than relying solely on outside suppliers.

At the core of Seagate's success is its advanced development of hard disc drive products. Seagate is the market leader in each of the segments in which it competes, ranging from price-sensitive desktops to performance-intensive network servers, and it produces a broad range of disc drives in capacities ranging from 20 Gbytes to an industry-leading 180 Gbytes. In the growing market of consumer electronics devices, which includes personal video recorder (PVR) products, gaming consoles and digital audio jukeboxes, Seagate has shipped nearly three million disc drives and assumed market leadership.

Seagate Technology is a global company employing nearly 50,000 people, with R&D and product sites in: the Silicon Valley, California; Pittsburgh, Pennsylvania; Longmont, Colorado; Bloomington and Shakopee, Minnesota; Oklahoma City, Oklahoma; Springtown, Northern Ireland; and Singapore. Manufacturing and customer service sites are located in: California; Colorado; Minnesota; Oklahoma; Northern Ireland; China; Indonesia; Malaysia; Mexico; Singapore; and Thailand.

5.3.2.2 Drivers for Change

Seagate is the world's leading provider of storage technology for Internet, business and consumer applications. Seagate's market leadership is based on delivering award-winning products, customer support and reliability to meet the world's growing demand for information storage.

Seven corporate objectives drive all day-to-day activities within the company.

They are:

- Improve time-to-market (TTM) for all products.
- Lead the industry in key technologies.
- Create world-class manufacturing operations.
- Develop strategic relationships with vendors and key customers.
- Provide best-in-class product and process quality.
- Become an employer of choice.

So why implement Six Sigma? The market leadership of the company is continuously challenged in a highly competitive and dynamic environment, as indicated by the following measures:

- Volume products remain in production only six to nine months.
- Technology content doubles every 12 months.
- Worldwide shipments of hard disc drives increase 10–20% per year.
- Cost per unit of storage drops 1% every year.

In 1998 Seagate's senior executive team was concerned that business performance was not on par with expectations and capabilities. The quality group was charged with recommending a new model or system with which to run the business. The Six Sigma methodology was selected and launched in 1998 to bring common tools, processes, language and statistical methodologies to Seagate as a means to design and develop robust products and processes. Six Sigma helps Seagate make data-based decisions that maximise customer and shareholder value, thus improving quality and customer satisfaction while providing bottom line savings.

Six Sigma was one of the three key activities seen as essential for Seagate's continuing prosperity. The other two were:

- Supply chain – how to respond to demand changes in a timely manner, execute to order commitments and provide flexibility to customers.
- Core teams – how to manage product development life cycle from research to volume manufacture.

For example, Seagate's Lean Manufacturing activities are a key part of Seagate's supply chain improvements and are increasingly tightly bound with Six Sigma. Lean Manufacturing's value stream mapping approach and Six Sigma's analytical strength fit together extremely well to define, solve and then prevent problems.

As Six Sigma matures at Seagate, leaders are shifting their focus from reactive to proactive deployment, and placing further emphasis on weaving Six Sigma into business areas in addition to operations. The company also deployed design for Six Sigma methodology (DFSS) – providing new tools and an emphasis on designing products based on a systems engineering approach. Thus, core teams are now starting to manage all drive, component and advanced development programmes using the DFSS methodology. The voice of the customer in the form of CTQ parameters is assessed against existing capabilities using a flow-up and flow-down process to identify any gaps that must be bridged to provide solutions to the customers' needs.

5.3.2.3 Implementation Design

The Six Sigma Academy was employed to guide the implementation and provide the initial waves of training for executives, champions and Black Belts throughout the latter period of 1998 and also 1999. Black Belt candidates were trained in the USA and Singapore. The three-year deployment plan followed the path of manufacturing operations first, then process and support engineering, followed by design engineers, administrative, sales and marketing, then began to engage suppliers and customers as well. Over the four training phases of DMAIC, each Black Belt candidate was expected to follow the 'learn–do' cycle with a real project being worked and reviewed both at the home site and in class. All sites were assigned Hands on Champions, members of senior staff familiar with the operational requirements of the sites and who were trained on project selection and support.

Seagate has now developed and customised training materials so as to be self-sufficient in educating up to the Master Black Belt level. Training centres of excellence exist in the USA, Europe and Asia-Pacific. Green Belt coaching is now required for all Seagate's professional and technical staff.

DFSS instruction was rolled out within design centres and functions, then through the advanced technology groups from 1999 to 2001.

5.3.2.4 Key Benefits Achieved

Go back to the year 1998 at Seagate. Upon hearing the term 'Six Sigma', the majority of employees probably stared blankly and asked, 'What in the world does that mean?' Four years later, and the snapshot looks completely different – Six Sigma, a household word at Seagate, is now highly visible and it is producing impressive results for the company, both in hard savings and improved business processes.

Looking first at these hard savings, Seagate has achieved nearly $700 million. All savings are validated independently and audited by the finance team, using very strict criteria. It is also certain that the so-called 'soft savings' that Seagate has achieved, but not counted, far exceed this value.

Hard savings, although an essential metric to track the progress of the activity, are only a small part of the story. Seagate's operational performance has improved tremendously on an overall basis. The company has moved into a technology leadership position that, in turn, has led to improved market share in an ever more demanding customer environment. While Six Sigma is not the sole contributor to this impressive performance, it has been an essential enabler. Six Sigma facilitates top line growth as well as bottom line savings.

As Six Sigma matures at Seagate, its leaders are shifting their focus from reactive to proactive measures. 'We've done quite well at cleaning up problems, so now we can evaluate and prevent potential problems', is one advantage according to Jeff Allen, vice president of Six Sigma and Design for Six Sigma. This leads directly into the increasing use of DFSS in order to leverage the effort earlier in the product life cycle.

Here are some measures of Seagate's progress:

- Over 600 Black Belts trained or in training – many now reintegrated into business functions after their two-year full-time assignment.
- Over 2700 Six Sigma projects completed.
- Green Belt education is nearly complete, with over 5500 employees trained.
- Over 1500 DFSS engineers trained or in training.
- Seagate's Six Sigma projects have delivered nearly $700m of cumulative, validated financial savings.

5.3.2.5 Lessons Learned

The main learning points from the Six Sigma programme at Seagate Technology include:

1. Companies using Six Sigma need to gain knowledge in how to use the metrics to manage – to make appropriate decisions on a holistic basis, avoiding sub-optimisation. This task of integration with the whole of the company's business process is the key.
2. Set aggressive goals – don't make them too easy.
3. Develop a system for tracking 'soft savings'.
4. Develop a common language and encourage its use on a widespread basis early in the programme.
5. Embed the business process within the organisation by training all functions – use Green Belt, Brown Belt and customised programmes as appropriate.

Company leaders are now asking, What is the next step for this programme? Are employees ready to take that step? And how do we get there?

In April 2002 a diverse group of representatives gathered for a two-day, intensive planning session to help define Seagate's future strategy for Six Sigma. The 14 representatives formed a collective vision for Six Sigma's future over the next three years and included specific elements that help to build a solid basis for future growth. Some of this input included:

- *Integration.* Seagate is at a point where the Six Sigma effort can start shifting from being viewed as an initiative to being integrated into the way the company does business. It becomes embedded within the organisation.
- *Metrics.* Critical processes will be well understood and their performance tracked with customer-focused, variation-based metrics. These metrics will guide company strategy and behaviour well beyond current implementation, taking Seagate to the next level of managing via metrics.
- *DFSS.* All products being developed will use the Design for Six Sigma process, enabling time-to-market and time-to-volume consistency and predictability. The DFSS process will integrate all functional groups into a global new product-delivery team.
- *Projects.* A dynamic process for prioritising projects and allocating resources will exist to verify that Six Sigma ventures are addressing current critical business issues.

Future plans include an ongoing assessment of the programme to make sure that Six Sigma is delivering exactly what it is intended to deliver.

Source: Rob Hardeman, Seagate Northern Ireland.

5.4 FITNESS FOR SUSTAINABILITY

The successful Six Sigma and Lean Sigma programmes, as illustrated by the above case studies, will be further aligned to the tenets of FIT SIGMA by adopting the four process components of 'fitness for sustainability':

- Performance management.
- Self-assessment and certification.
- Senior management review or sales and operational planning (S&OP).
- Knowledge management.

It is likely, although not verified, that both General Electric and Seagate are following in some form the above processes. The large manufacturing organisations engaged in major Lean Sigma programmes should also ensure that these four processes of FIT SIGMA are implemented to sustain the results and a continuous improvement culture.

5.5 FIT SIGMA IN SUPPLY CHAIN MANAGEMENT

Supply chain management in a large manufacturing and supply organisation considers demand, supply and inventory needs for each item of production. In the 1960s and 1970s the manufacturing and supply strategy of multinational companies focused on vertical integration. In the 1980s and 1990s large organisations started to concentrate on their 'core businesses' and divested other arms of the trade. The beginning of the 21st century is now witnessing the explosion of outsourcing and emergence of competent but lower-cost manufacturing in Eastern Europe, China, India, Brazil and South East Asia. A little more than a decade ago the companies were urged to attain so-called 'world-class' performance within the enterprise. The departments within a company were striving for islands of superb quality; then, with a succession of operational excellence initiatives (e.g. TQM and Six Sigma), the fences between departmental turfs were gradually demolished.

New developments in supply chain management clearly indicate that focusing on the conventional practices of supply chain management within the organisation may achieve operational excellence only inside the business's own infrastructure. The benefits? Well, it would be like driving a trouble-free high-performance motor car in a traffic jam, or it could be comparable to having a perfect stainless steel link in a rusty chain.

The recall of Toyota cars in March 2010 illustrates the above point. Toyota motor cars have been the global market leader, both in quantity and quality. The Toyota production system has been the gold standard of quality and Lean Manufacturing, both in the academic and business spheres. The recall of millions of cars mainly due to the 'sticky pedals' on the Prius has surprised the quality experts, although recalls in the motor industry for quality and safety reasons are quite frequent. The impact was so serious that the top boss of the company, Mr Akio Toyoda, had to formally apologise to the US Congress on 24 March 2010. So what went wrong in this iconic company?

The primary reason for the recalls is that Toyota kept most of its development in Japan, although it built research and development centres in places like Ann Arbor, Michigan. Despite its global expansion during the 1990s, the company failed to hire enough engineers to keep up with production increases and there was a restriction on overtime regulated by a new Japanese law. Mr Toyoda admitted that the rapid expansion of the company could not cope with the attendant safety issues.

Secondly, because of the global expansion and outsourcing, it was difficult for Toyota to train and sustain the legendary quality assurance programmes with all suppliers.

Another issue is that Toyota, like other global auto companies, has outsourced the development of crucial components to its suppliers. As a result, for example, the trouble with the Prius pedal was caused by a software problem. Furthermore, there was difficulty with an interaction between the US supplier of material for one part of the pedal and Denso, Toyota's Japanese supplier of accelerator pedals.

This is a good learning experience for all of us and it emphasises the argument that an operational excellence initiative (such as FIT SIGMA) of a large manufacturing organisation should be shared with key suppliers and business partners of the total supply chain.

The following case example illustrates how a FIT SIGMA initiative can be shared with partners of the total supply chain.

Case example: FIT SIGMA in the caterpillar supply chain

Caterpillar, or Cat, is the world's largest manufacturer of construction and mining equipment, diesel and natural gas engines, as well as industrial gas turbines. They are very well known and most consumers would recognise their distinctive huge yellow equipment. It also delivers many related services, including financial, logistics and remanufacturing. Caterpillar reaches every continent with its products and services and, as of 2008, employs approximately 113,000 people. Caterpillar typically doesn't sell to the end customer but uses a global dealer distribution network. As of 2008, 180 dealerships span North America, Europe, Africa, the Middle East, Asia and Latin America. These dealerships, which have more than 131,000 employees, act as the contact point between the end customers and Cat.

Caterpillar CEO Glen Barton started the Six Sigma culture change in 2001 when he challenged every manager to show his or her personal commitment to the programme.

Since then, the Six Sigma culture has permeated Caterpillar in all aspects of the business.

In addition to its own use of Six Sigma, the company has taught its suppliers and dealers about the benefits of using the technique to refine the entire sales model. Caterpillar has introduced 850 suppliers world-wide to Six Sigma, which has created more than 1000 supplier Black Belts to help run the projects. One supplier that said it was interested in the Caterpillar Six Sigma methodology allowed Cat to consult and transform the business. When implementing Six Sigma, Caterpillar used facts and data to show the results the supplier could expect, so it didn't take long for the supplier to be persuaded and to 'buy in' completely to the methodology.

Dealers have also taken on the Six Sigma commitment. More than 165 dealerships have produced over 1000 Black Belts to help with projects. Dealers find it amazing that they can share their projects with one another on a Caterpillar website that depicts best practices among the dealers. Even though each dealership is run as a separate business, Six Sigma has helped give all of them a common feel across the world. Not only are dealerships learning about projects that need to be done in their business, but they're following the steps of the process and learning which projects to tackle first. Just as Caterpillar embraced the methodology, dealers have also accepted the idea of making Six Sigma a top-down methodology that pushes the training and concept down to the workers at the lowest level.

Source: Strategic Finance, April 2010, pp. 25–28.

5.6 SUMMARY

With the dynamic growth of emerging new economies (e.g. China, India, Eastern Europe and Latin America) and the globalisation of manufacturing businesses, in order to remain competitive manufacturing companies need to implement continuous improvement which achieves sustainable changes. Many large multinational manufacturing companies may have lost their appetite for Six Sigma, declaring 'We have done it and got our T-shirts'. The 'fitness for sustainability' approach of FIT SIGMA should help rekindle enthusiasm in these organisations and address any lethargy regarding the process.

Many large manufacturing companies may have experienced cumulative implementation problems. Other businesses who have not yet started Six Sigma may be questioning the value and cost-effectiveness of the commitment of considerable resources to implement Six Sigma. The 'fitness for the purpose' tenet of FIT SIGMA should help this group of large manufacturing companies.

The 'Sigma (Σ) for improvement and integration' approach of FIT SIGMA for large manufacturing organisations advocates the deployment of DMAIC Full. This should be supported by comprehensive training, including the development of both Black Belts and Green Belts.

FIT SIGMA tips for large manufacturing organisations

1. Assess where you are in quality management and continuous improvement initiatives and align with business strategy and objectives.
2. Perform a preliminary stakeholder analysis and identify the Champion or Torch Bearer.
3. Carry out a value stream mapping for key products and start with Lean to eliminate waste.
4. Start with the division with most interest and influence in the programme.
5. Select consultants/trainers for Black Belt, Green Belt and Champions training.
6. Deploy DMAIC Full methodology.
7. Use FIT SIGMA tools for all projects and subsequently advanced statistical techniques only for specific complex and critical problems.
8. Execute and review projects and monitor costs and benefits.
9. Develop Master Black Belts and in-house capability and roll out to all divisions, including major suppliers in the supply chain.
10. Sustain the initiative and culture by performance management, senior management review, self-assessment and knowledge management.

6 FIT SIGMA in Service Operations

6.1 INTRODUCTION

It is generally agreed that 75% of the workforce in the United Kingdom is engaged in service industries. This high proportion is not unique to the United Kingdom; indeed, it is representative of employment statistics for developed nations throughout the world. In fact, the US Census Bureau shows that over 80% of the workforce in the United States is employed in service industries. There are many service businesses such as healthcare, banking, food services and retailing. In addition, there exist numerous service functions in both manufacturing and service organisations, for example marketing, finance, information technology, procurement, legal and human resources. Although a shift back to manufacturing industries has been identified (Basu and Wright, 1998), nonetheless it is obvious that the larger percentage of the workforce of developed nations will continue to be employed in service type activities. There are several reasons for this.

Firstly, continual advances in technology mean that manufacturing is considerably less labour-intensive than during previous times. Automation, robotics, advanced information technology, new materials and improved work methods all have led to the decimation of manual labour.

Secondly, for larger organisations, manufacturing has become internationalised. Many companies (for example Nike) that began with a manufacturing base are now seen as primarily marketing and service companies, with manufacturing being supplied by contractors or allied companies situated all round the world.

Additionally, organisations can no longer regard themselves as being purely in manufacturing and hope to survive. The market first and foremost now demands quality of product *and* service. Market expectations of the level of quality are driven by perceptions of what technology is promising and by perceptions of what the competition is offering.

Organisations now operate in a global forum where national barriers and tariff and customs duties no longer provide protection for a home market. Any manufacturer, even if it has concentrated its efforts on supplying a local market, is in reality contesting on the world stage. Competition is no longer limited to other local organisations and the fiercest battle in the home arena will be from overseas produced goods. This fact alone has meant that manufacturers can no longer make what suits their engineering strengths, but must now be aware of what the market wants and what global competition is offering. And what the competition is offering is service, in the form of delivery on time, marketing advice, training, installation, project management, or whatever else is required to provide a total service as well as a reliable product. For example IBM, the consistent leader in the development of information technological for over 50 years, now no longer relies on selling technology, but rather upon market solutions. This means getting alongside the customer, finding their problems and then using the technology to provide a solution, with the emphasis being on the solution and not the technology.

Finally, it has to be recognised that customers today are better travelled, more informed and have higher expectations than did their predecessors. A great deal of customer expectation for a continuously improved product and service has arisen from global competition, and the well-publicised TQM drive of the 1980s in service industries. Thus, for over 20 years customers have been given the message that they are of almost regal importance, and have the right to expect correct products that work first time as well as a high level of service. Even government departments have mission statements espousing service, although the reality might be somewhat less than that desired by the customer!

Consumers therefore expect and take for granted that they will receive a reliable, high-quality product for their money. Most organisations realise that their products actually differ very little from those of the competition, and that any technological improvement is soon copied, so that the only real difference – the 'competitive edge' – comes from service.

6.2 THE DIVIDE BETWEEN SERVICE AND MANUFACTURING

If no serious operation can ignore market demands for service and world-class quality then why bother to try and separate manufacturing from service? Indeed, for a manufacturing organisation aspiring to world-class status we would agree, most emphatically, that the management of such organisations must concern themselves with service and quality if they are to compete on the world stage. But let us consider a manager in a service industry such as health, retail, distribution, education, travel, real estate, consultation, brokering, law, accounting, administration of central and local government, transportation of goods or people . . . the list is endless. This manager works in an area where no direct manufacturing is involved, or where the manufacturing is light and simple (such as in a restaurant), and actually does not have to know much about manufacturing. Although all the above industries are reliant on manufacturers to varying degrees for the equipment they use, or in the case of a retailer for the goods they sell, the actual physically heavy work of making the goods is not their concern. The analogy is that of a driver of a car: one can be a very good driver without knowing very much about what happens under the bonnet. Some basic knowledge, such as realising when to change gear and knowing about the danger of overheating due to lack of oil or water will be of advantage, but not much more is really necessary. Likewise for management in service industries, a detailed understanding of line balancing for a high tech mass production assembly line of washing machines is not necessary to be a retailer of white wear. Some familiarity with lead times for deliveries, operating instructions and the capacity of the washing machine will be sufficient for the sales person as a basis for dispensing good advice and service to the customer.

Thus, there can be a separation of management into two distinct streams. One is the management of production operations including service, and the other is the management of operations in service industries where only some rudimentary knowledge of manufacturing is required.

Regarding the first stream, the manager that is involved directly in production and manufacturing needs to be well versed in the strategies, tactics and methodologies of production management and also has to be very aware of what constitutes service and quality from the customer's point of view so as to be able to provide a quality product coupled with the service required to better the competition.

The second stream consists of managers who are primarily engaged in service industries. These people do not require a detailed knowledge of production systems and methodologies. Instead, they need to concentrate on providing a better service than the competition, and using their resources as efficiently as possible to provide this high level of assistance.

6.3 OBJECTIVES OF A SERVICE ORGANISATION

A service organisation exists to *interact* with customers and to satisfy customer service requirements.

A service organisation is when two or more people are engaged in a systematic effort to provide services to a customer, the *objective* being to *serve* a customer. For any service to be provided there has to be a customer. Without the existence of such a customer, and *interaction* between the customer and the service organisation, the objective of providing service cannot happen. The degree of intensity of interaction between the customer and people of a service organisation varies and depends on the type of service offered. For example, a specialist medical consultant will have a high degree of 'face-to-face' interaction with the customer, and likewise so will a hairdresser.

Further down the scale of face-to-face interaction is a restaurant where the customer will judge the quality of service by the level of interaction (knowledge of the wine waiter, attentiveness of waiting staff, etc.), as well as by the standard of the goods provided (wine and food). The restaurant in turn will, however, have a higher degree of personal interaction than will a fast food takeaway. Irrespective of the level of 'face-to-face' interaction, without some customer interaction service cannot be provided. Note, this does not mean that the customer always has to be present when the service is being provided. For example, when a car is being worked on by a mechanic, the owner of the car need not be present, but nonetheless without the owner actually supplying the car and giving instructions ('grease and oil change, ready at five?') no service can be provided. Most types of operation are a mixture of both products and services, as shown in Figure 6.1.

Delivery on time is an obvious measure of performance, but measurement of the effective use of time and information is less easy. The attitude and reliability of people is a value judgement, which is even harder to measure. Nonetheless, the amount of time and information available will be important issues for the manager of an operation, and likewise the friendliness and helpfulness of staff will be important issues for the customer.

With today's technology, information would seem to be readily available. The concern is *knowing* what information is required, and then being able to *interpret* and *use* that information so as to achieve the organisation's operational objectives.

When introducing the concept of customer satisfaction, it has to be understood that the basic requirement for the customer is that the service must firstly match their specification and secondly meet time and cost constraints.

Specification – providing the customer with what they expect to receive or are prepared to accept – is the essential prerequisite.

Time and cost are the next requirements. The service has to be provided at a time that is acceptable to the customer and the price must be reasonable.

Businesses in the service sector have often questioned the tangible value that a Six Sigma programme delivers in services where there are many intangible and transactional processes. Historically, the first organisations to adopt Six Sigma were mainly in the manufacturing sector. This was due to the fact that the core Six Sigma methodology revolved around the

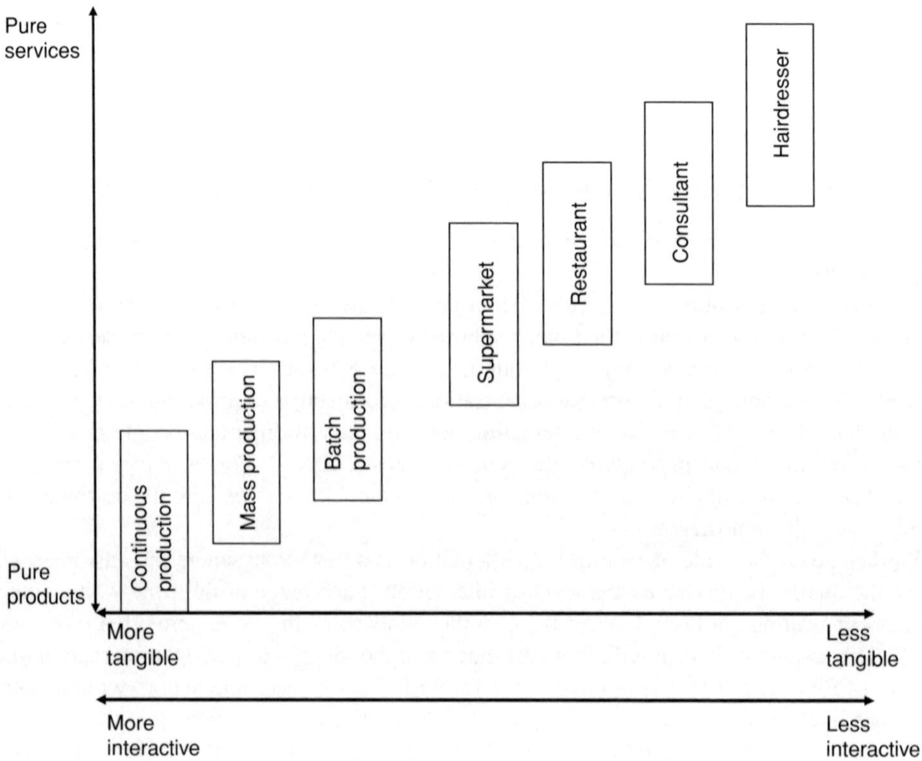

Figure 6.1 Operations showing level service content and interaction

reduction of defects in a process. As with aircraft engines, this might be a fault in a component that could typically lead to a catastrophic failure. Hence, a robust quantitative methodology of a very high confidence level such as Six Sigma lends itself well to the prevention of problems in this type of scenario. Services, by their nature, are very often bound by time in terms of the processes that are run and lead to the delivery of an outcome that then benefits a customer. The confidence level in services is likely to be lower than an aircraft engine in this scenario, although the customer expects a consistent service level. This is where FIT SIGMA comes in as a methodology that looks at how waste (in terms of time) may be taken out of a process and allows that process to become more efficient and, in turn, build a sustainable service to the customer.

The level of service offered stems from the business policy, which in turn is, to a large extent, driven by what the competition is doing, or is threatening to do. When deciding on the level of service, and specifying that level of service, management to a large extent relies on the advice of the marketing function. If the marketing function does not correctly interpret the requirements of the customer then there will be a *gap* between the level of satisfaction the organisation believes it is providing, and what the customer believes they are getting. The concept of 'service gaps' arose from the research of Zeithaml *et al.* (1990). As Lewis (1994, p. 237) says, 'They defined service quality to be a function of the gap between consumers' expectations of a service and their perceptions of the actual service delivery by an organisation;

and suggested that this gap is influenced by a number of other gaps which may occur in an organisation'.

The magnitude of the gap will be compounded by the number of steps in the service process and by the distance of the operational function from the customer. Unless gap analysis is attempted, management will firmly believe that the overall result is somewhere near 90% of what the customer wants. And each department when queried will fervently believe that they are reaching between 90% and 95% of required performance levels.

If an organisation is close to its customers and aware of what the competition is doing, then a gap of this magnitude should not happen. Two factors are at play here. Firstly, the larger the organisation and the greater the delineation of responsibilities between departmental functions, and secondly, the further the operations function is removed from the customer and from consultation in business policy decisions, the greater the likelihood of gaps occurring between what is provided and what the customer really wants. The integration of processes between departments is also an objective of FIT Σ.

6.4 'FITNESS FOR THE PURPOSE' FOR SERVICE ORGANISATIONS

So if the process characteristics and management tools of service organisations overlap with those of manufacturing, is it not logical to assume that the FIT SIGMA approach for both sectors should be the same? The broad answer is yes. The 'fitness for the purpose' methodology of FIT SIGMA can accommodate the variability between service and manufacturing as much as the variability with the manufacturing sectors. However, there are areas where some differentiation may be applicable, in particular, in 'service level' and 'culture'.

Although it varies depending on the particular function, the process σ governing the service level of a service organisation is likely to be less than a zero-defect manufacturing organisation. An error in a service function may cause severe financial penalty, but a defect in a manufactured component may cause loss of lives. A change from 6σ to 5σ is equivalent to unsafe drinking water for 15 minutes every day. Table 6.1 gives an indication of how many 'process sigmas' may be appropriate for most of the service functions.

Table 6.1 Process sigma targets

Service	Process σ	Manufacturing	DPMO
—	6	World Class	3.4
World Class	5	Excellent	233
Excellent	4	Good	6,210
Good	3	Average	55,800
Average	2	Poor	309,000
Poor	1	—	69,100

The following case example illustrates how a Lean Sigma approach conforms to FIT SIGMA in a service organisation such as an accountancy firm. In this scenario, the objective was to reduce waste and increase efficiency with a confidence level of 90%. The start-up cost was not significant as the firm initially did not deploy an extensive training programme for Black Belts and Green Belts.

Case example: FIT SIGMA in an accountancy firm

Rea & Associates, Inc. is an Ohio-based accountancy firm with 11 locations and 250 employees. The firm started a Lean Sigma programme in 2007 by recruiting a Six Sigma Master Black Belt. Initially, with about a year of start-up costs, including tuition for the additional Black Belts and Master Black Belts, the programme was slowly rolled out as a Lean Six Sigma tax programme. The approach of Lean Sigma conforms to FIT SIGMA in this application.

Much of the waste in tax processes was viewed as being small in nature and therefore was overlooked. Nearly every step of a process had some form of waste. It was important to identify this, however small, and to quantify it to show the entire impact to the procedure. 'Minor' wastes added up to 30% of efficiency with a confidence level of 90%. Table 6.2 shows the results in Autumn 2008.

Table 6.2

Lean metrics:	From	To
Waiting time for review	Two-week average	Less than one week
Waiting time to clear review notes	One-week average	Less than two days

The same principles discussed for the tax preparation function also applied to audit, payroll, bookkeeping and pension administration, to name a few. These ethics applied if the desire to improve efficiency and quality remained competitive. This approach also worked well when applied to internal administrative tasks, such as billing. Regardless of the internal service being performed, this method was also used to analyse and improve the process.

Source: Adapted from the *Journal of Accountancy* 2010.

'Fitness for the purpose' in the FIT SIGMA methodology also conforms to the application of Six Sigma and Lean Sigma in large service organisations such as GE Capital, American Express and the Bank of America. In this scenario the organisations can afford a large budget for a comprehensive deployment of the training programme for Black Belts and Green Belts.

6.5 'SIGMA (Σ) FOR IMPROVEMENT' FOR SERVICE ORGANISATIONS

Snee and Hoerl (2009) recommend that 'transactional or service processes such as accounting and logistics, and non-many manufacturing organisations such as hospitals and banks can benefit from Lean Sigma, but it is important to understand the difference between service and factory applications in order to optimize this win–win tool'.

The FIT SIGMA improvement approach conforms to the above recommendations. There are technical differences between a service and a manufacturing operation. The first is the lack of suitable measurement systems to obtain accurate data. Another discrepancy in a service organisation relative to manufacturing is that many service processes are not standardised (although it can be argued that services like accounting are standardised before they are computerised). However, the main point of variation is cultural and it is due to the fact that service businesses and functions typically lack engineers with a scientific approach.

Table 6.3 FIT SIGMA for manufacturing and service organisations

	Manufacturing	Service
Programme start-up	Easier to convince top management	Difficult to convince top management
Appropriate tools	SIPOC, flow process, DOE, fishbone, Pareto, DMAIC DFSS	Same Questionable
Learning deployment	Structured training programme and process owners	Same
Project selection	Easier to quantify savings	Difficult to quantify savings
Implementation	Likely to have experienced 'derailers' and thus more difficult to implement	Easier to implement after top management approval
Sustainability	Senior management review Knowledge management Performance management Self-assessment	Same Specifically adapted for service

This cultural difference is also rooted in the fact that service organisations are in an early stage of the learning curve of applying tools and techniques that have been implemented in manufacturing many years back. This gap in experience should be recognised and reflected in a FIT SIGMA programme, as we have summarised in Table 6.3.

In large service organisations, such as banks, hospitals and supermarkets, a DMAIC Full approach (see Figure 6.2) is appropriate.

The following case example in a large health service organisation illustrates how a DMAIC Full approach is applied to integrate all departments and improve processes.

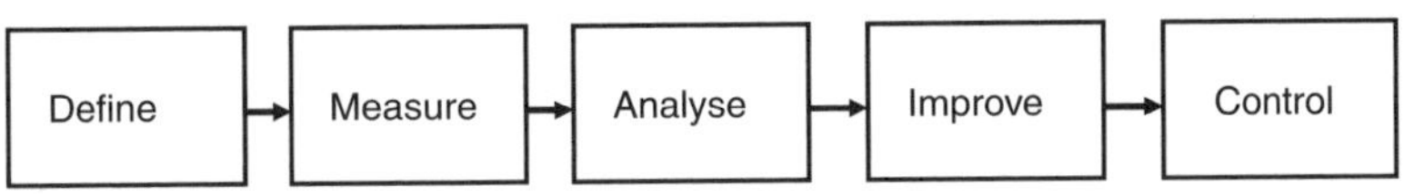

Figure 6.2 DMAIC Full approach

Case example: FIT SIGMA in a community hospital

Floyd Medical Center, a stand-alone safety net community hospital, is located in northwest Georgia, USA. Floyd engaged the Health Management Corporation (HMC) to benchmark the organisation in the early 2000s and began its journey into Lean Six Sigma by investigating what measurement and improvement methodologies other organisations were using. Floyd's initiative began in the summer of 2006 with a management retreat with Chip Caldwell and Associates, who were engaged to assist the initiative and remain for a year to conduct training. After the retreat, leaders were educated about the cost of poor quality and the concept of driving out waste to improve quality, cost, time and performance, and the change model of the 100-day workout was introduced. These 'workouts' are comparable to FIT SIGMA training modules for Black Belts and Green Belts.

The first 100-day workout was a waste walk. Future 100-day workouts utilised the Suppliers, Inputs, Process, Outputs, Customers (SIPOC) tool, the Define, Measure, Analyse,

Improve, Control (DMAIC) process, and other Six Sigma tools. At the 100-day mark, a summation of all the changes and savings was reported and the next workout was kicked off. This exercise was used to train and certify the first group of Green Belts and Black Belts. Four employees continued to obtain Black Belt certification in August 2007 and 12 employees were trained as Green Belts.

In this first workout 288 changes were instituted and 270 were validated, for a savings of $2,458,200. In storage costs alone, $63,500 was saved. 'The Greatest Financial Impact' award was given for the recommendation to switch Zofran to a generic drug. This yielded a savings of $223,800.

The second workout dealt with waste and in-quality staffing. Results of this study were reductions in overtime, decreased utilisation of agency staff and a reduced number of positions through attrition. In total, 145 changes were instituted, 131 were validated and a savings of $1,543,600 was achieved.

The third workout concerned waste and SIPOC. Leaders identified flow and throughput issues that could be improved. This exercise made leaders think about the customers of each step in a process. This workout resulted in a savings of $635,895. The next three workouts also produced excellent outcomes, more tools were implemented and this also led to a refined reporting process. However, after the sixth workout, the management felt that a change was needed to sustain the momentum and the level of performance.

Lessons learned

There are several important lessons learned from this case example.

First, the initiative must be driven from the top down. Second, knowledge and training in the methodology are critical at all levels of the organisation. Third, it is necessary to identify a Black Belt to work on projects full time. Belt duties were often carried out after regular work hours, and it was found increasingly difficult to have Belts available to meet project needs and grow the initiatives. Therefore, one Black Belt was identified to focus full time on such ideas. Fourth, it is essential to limit the number of changes and efforts in a particular area at any one time. You are likely to set them up for failure with too many efforts running in tandem. Finally, there is a need to have processes in place to sustain the enthusiasm of the team and the performance level. These learning points – and in particular the final one – conform to the tenets of FIT SIGMA.

Source: *Frontiers of Health Services Management*, Volume 26, No 1.

6.6 FITNESS FOR SUSTAINABILITY

As described in Chapter 2, the fitness for the sustainability of performance in FIT SIGMA is underpinned by four key processes:

- performance management;
- senior management review;
- self-assessment and certification;
- knowledge management.

The continuing case example of Floyd Medical Center illustrates how focusing on performance management and knowledge management (e.g. continuous training) helped the programme to sustain momentum and the level of performance.

Case example: Sustainable focus in a community hospital

At the end of the sixth 'workout' or workshop, the Lean Sigma programme at Floyd Medical Center in Georgia began to lose momentum. Leaders were becoming weary. While the first wave had been very successful, with significant savings in each workout, the management decided to change focus. Some years earlier, Floyd had adopted a Balanced Scorecard we call the 'Value Compass', comprising five aspects (e.g. Financial, People, Strategy, Customer Satisfaction and Internal Quality). This Value Compass became the focal point of subsequent workouts.

For the seventh workout, called the Fall Value Compass Workout, seven teams were chartered to enhance the Value Compass in seven areas. These were:

1. Emergency care center throughput (Strategy).
2. Orthopedic and neurology service line (Strategy).
3. Patient satisfaction (Customer Satisfaction).
4. Employee choice (People).
5. Hygiene and antibiotic stewardship (Internal Quality).
6. Revenue cycle (Financial).
7. Time management (People).

The above Value Compass focused on projects in the seventh workout and also applied DMAIC as well as contributing savings of \$668,000. The focus on a Balanced Scorecard-based performance management conforms to the FIT SIGMA tenet of sustaining performance. Floyd Medical Center is committed to 'workouts' based on the Value Compass and also aiming towards the following conditions of sustainability:

- Requiring all departments to have control charts.
- Having Black Belts serve as full-time Belts and then rotating them to other leadership positions.
- Requiring all directors to obtain Green Belt certification, and making that a prerequisite for anyone who wants to advance into a director position.
- Having more nurses trained as Green Belts.
- Having one or two Master Black Belts within the organisation.
- Establishing a network of similarly minded hospitals throughout the country who want to share their Lean Six Sigma experiences.

Source: Frontiers of Health Services Management, Volume 26, No 1.

The approach of Floyd Medical Center towards the sustainability of the programme is on the right track in its efforts to conform to FIT SIGMA. The healthcare centre will benefit further by adopting a regular self-assessment process.

6.7 SUMMARY

In this chapter we have determined and demonstrated by case examples how FIT SIGMA is applied in service organisations. The appropriateness of tools and methodology should be balanced against the available resources, the cost effectiveness of the programme and the level of service required for the organisation. The determinant of the level of service to be provided will be driven by the competition and demands of customers and stakeholders. To provide the necessary *affordable* level of service, management are vitally concerned with the efficient and effective use of resources. Such assets will generally be limited in quantity and quality. Therefore, there are potentially conflicting objectives: customer satisfaction and efficient resource utilisation. FIT SIGMA recognises this and provides an approach to determine what the customer really wants, and how to match resources to essential customer needs. With FIT SIGMA it is shown that it is not essential to meet all customer requirements, but by making the best of existing resources it is possible to meet key needs. In Lean Sigma the quality of customer service (the voice of the customer) is addressed by Six Sigma and the utilisation of resources is balanced by Lean. FIT SIGMA further enhances Lean Sigma with 'fitness for the purpose' and 'fitness for sustainability'.

FIT SIGMA tips for service organisations

1. Assess where you are in quality management and continuous improvement initiatives and align with business strategy and objectives.
2. Perform a preliminary stakeholder analysis and identify the Champion or Torch Bearer.
3. Carry out value stream mapping for key products and start with Lean to eliminate waste.
4. Start with the division with most interest and influence in the programme.
5. Select consultants/trainers for Black Belt, Green Belt and Champions training.
6. Deploy DMAIC Full methodology or DMAIC Lite methodology depending on process sigma targets.
7. Use FIT SIGMA tools for all projects.
8. Execute and review projects and monitor costs and benefits.
9. Develop Master Black Belts and in-house capability and roll out to all divisions.
10. Sustain the initiative and culture by performance management, senior management review, self-assessment and knowledge management.

7

FIT SIGMA in Small and Medium Enterprises

7.1 INTRODUCTION

Small and medium enterprises (also known as SMEs) are companies whose headcount or turnover falls below certain limits. As shown in Table 7.1, an enterprise with a turnover of less than €50 million and employing fewer than 250 people is classified as an SME.

SMEs are underpinning the economy of a country, accounting for 50% of the annual turnover in the private sector in the United Kingdom (Berr, 2008). One of the more familiar challenges in business today is how to implement Six Sigma in small and medium-sized companies. This trial is compounded when larger companies are beginning to demand Six Sigma applications to their supply base, comprising SMEs as a condition of conducting future business. However, when the SMEs solicit advice and proposals from Six Sigma consulting companies, they learn that the traditional Six Sigma implementation approach can require a disproportionate investment and the dedication of their best full-time resources. An empirical study by Antony *et al.* (2005) has also found that many of the SMEs are not aware of Six Sigma and several do not even have the resources to implement Six Sigma projects. The authors also indicate that there is currently insufficient research into the effects of Six Sigma on SMEs, making it very difficult for any concrete conclusions to be drawn.

Harry and Crawford (2004) refer to two new ranks within the Six Sigma belt system and identify these as White and Yellow Belts. These are in addition to Master Black Belts, Black Belts and Green Belts. According to the authors, a White Belt can offer a much quicker return on investment, by completing approximately one project per month, rather than the two to three a year that is expected from a Black Belt. There is no evidence that this new system will actually work with SMEs. However, this does indicate that the original Six Sigma approach, the 'one size fits all' philosophy, in actual fact might be flawed. It is therefore important to note that an appropriate level of training is key to the success of Lean Sigma in SMEs. A recent study by Deshmukh and Lakhe (2009) supports that the training forms an important factor to achieve the desired effect of a Six Sigma implementation programme in SMEs and suggests that the DMAIC model needs to be customised and enriched with T (training), thus making a customised process of 'T-DMAIC' for SMEs. FIT SIGMA emphasises the appropriate education and processes for SMEs rather than the 'one size fits all' approach.

Brenig-Jones (2008) proposes 'Kai Sigma' as a combination of Kaizen and Six Sigma methodology applied more intensively over a shorter time horizon. It is intended to provide both a systematic and pragmatic way for people and organisations to improve performance. This approach is likely to be acceptable in an SME. The manager of Kai Sigma in an SME can bring together the concepts of Kaizen to involve people in continuously seeking to improve performance within the framework of DMAIC.

There are examples of many successful applications of the Six Sigma methodology in SMEs in the United Kingdom (Davies, 2005), Germany (Wessel and Butcher, 2004), Denmark (Barfod, 2007), India (Kumar *et al.*, 2006) and Indonesia (Achanya *et al.*, 2006).

Table 7.1 SME definition

Enterprise category	Headcount	Annual turnover	Balance sheet
Medium	< 250	< €50 million	< €43 million
Small	< 50	< €10 million	< €10 million
Micro	< 10	< €2 million	< €2 million

Source: European Commission (2003).

The findings indicate that the results are quicker and more visible in smaller companies than in larger corporations. However, the DMAIC approach had to be modified to be applicable and valuable in the SME environment. The following case example to retain customers in a call centre illustrates this point.

Case example: FIT SIGMA to solve a call centre problem

Six Sigma methodologies can be tailored to be as simple or as complex as you want them to be, dependent on the project the business has in mind. This approach then conforms to FIT SIGMA.

A customer relations project in a call centre demonstrates how problems that are perceived to be at the 'end' of a process are actually shown to be issues that have arisen earlier on.

The company was experiencing problems within the customer call centre. It was inundated with calls, most of them problems. Significant money had been spent on extra staff, new training, looking at limiting the time taken per customer call, etc. The call centre was deemed crucial as the business had a 50% customer turnover annually, and had made no real effort to focus on and improve customer retention. The business then decided to undertake a Six Sigma project to look at the problem, which began by identifying who the call centre was actually trying to serve – the company or the customer!

Initial analysis captured how their customers defined 'value', and one of the key CTQs was 'customer focus'. The key attributes of 'customer focus' were responsiveness, consistency, problem resolution, ease of dealing with the call centre and an understanding of customer needs.

These key elements all scored poorly when applied to the company. The business then used this data to map out the 'order to delivery' process and identify areas where improvements could be made. Interestingly, whilst many of the problems were within the call centre, some difficulties occurred earlier in the sales process. The root cause analysis indicated that:

- Sales people were being paid commission on the number of new customers, leading to incomplete applications and generating more calls from customers to the call centre.
- Customers were not being trained in how to use the product properly at the time of the sale, leading to more calls to the call centre.

The solution had three parts:

1. The business introduced an electronic order form that could not be submitted until all field processes had been completed.

2. Sales team job descriptions were changed to make sure that customer training was emphasised.
3. Call centre personnel were offered training on how to retain customers.

Outcome – customer turnover reduced from 50% to 25%.

As you can see from the above example, all that was needed to resolve the call centre issue was an examination of the process that the customer was subject to and to fix any areas that were deemed as problematic to the customer. This did not involve any extra cost or any staff redundancies, and customer retention was greatly improved.

This example shows that Six Sigma can be used effectively in the SME marketplace, and it can be made simple and relevant to many sectors and functions. This is FIT SIGMA in action.

Source: sixsigmaiq.com.

In the current economic downturn enterprises all over the world, including SMEs, are forced to do more with less and to trim costs while growing profits. It is likely that the reduction of waste by Lean processes is more attractive to SMEs than a commitment to a Six Sigma programme. In this scenario the appropriateness of the three tenets of FIT SIGMA becomes more relevant for SMEs.

7.2 'FITNESS FOR THE PURPOSE' FOR SMALL AND MEDIUM ENTERPRISES

Many SMEs have dived into a Six Sigma programme without going through the earlier crucial stages of identifying the real requirements. Arguably, a significant number of companies that initiated a Six Sigma programme did so because they felt threatened in terms of their very survival, or they became victims of management fads. In some cases Six Sigma was attempted as a panic measure when it was too late, and disappointment was expressed when the bottom line and share price did not miraculously improve within a short period. As discussed in Chapter 2, it is advantageous, although not essential for SMEs, to apply the initial assessment process based on a set of questions. These could be employed later for the periodic self-assessment or for a certification process.

SMEs can speed up the structuring of an effective FIT SIGMA programme more than large businesses. Examples might include flexible process flows, a short decision-making chain and the high visibility of senior management, etc. On the other hand, there are factors that can be disadvantageous, such as a lack of resources and expertise in change initiatives. Therefore, many SMEs have achieved success in Lean Sigma initiatives when they are led by senior managers. This approach conforms to the tenet of 'fitness for the purpose' in FIT SIGMA.

In an SME, due to the limited number of high-calibre resources, some senior managers can be trained as Green Belts and may then lead projects. FIT SIGMA undertakings in an SME should be derived from the business' strategic plan, which sets an organisation's goals and key performance measurements. Based on key business goals, senior management defines the scope of each FIT SIGMA venture and organises the right project team. While each assignment is progressing, they regularly carry out reviews and swiftly provide the appropriate political, financial and technical support. After all tasks are completed, senior management audits the project's results, establishes necessary systems to sustain the improvements, and continuously adjusts the business strategic plans.

The 'fitness for the purpose' philosophy applies to the type or size of business, whether it is manufacturing operations in a factory or a service procedure in an office. The success of GE Capital after the application of Six Sigma is well documented (Welch, 2001, p. 231). However, FIT Σ methodology is also applicable to a relatively small service operation, as the following case example of a legal practice will illustrate.

Case example: SIX SIGMA in a legal practice

Countrywide Property Lawyers (CPL) is a subsidiary of the Countrywide Assured Group in the UK and was established in 1997 to provide a specialised conveyancing service on residential property. Five CPL centres were opened in Woking, Northampton, Brentwood, Cardiff and Manchester. CPL has fewer than 400 employees but it is the largest conveyancing company in the UK.

Conveyancing is fast becoming a commodity product that can be bought 'off the shelf'. Flexibility, the use of IT enablers and quality customer service are the differentiating factors in a competitive market worth £400 million in England and Wales.

Part of the current conveyancing process is to examine the quality of the data and information, including the review of contracts, title deeds and financial information. Currently CPL does not know if their conveyancing process is under control. Client satisfaction questionnaires which are filled in by customers show a large variation in the conveyancing lead time of between 8 and 15 weeks. Although CPL currently reports that on average it takes 10 weeks to see a transaction through from 'instruction to completion', this figure is proving to be misleading due to the large variations in the process.

CPL considered the application of the Six Sigma concept to reduce variation in lead time. The initial assessment indicated that more than 50% of transactions take longer than 10 weeks and more than 15% take over 14 weeks.

This is an example where the FIT Σ methodology can be applied. The focus will be to train five experts (one on each site) specifically on the connection between averages and span in variation. These experts will train other employees and thus variation control will be built into the conveyancing process by employing the FIT Σ methodology. The target for 'span' will change from 8–15 weeks to 9–10 weeks. The losses will be minimised and customers would be able to make, for example, firm arrangements for a removal van. This is obviously advantageous, as anyone who has experienced the stressful and frustrating 'limbo' between a house sale and legal completion will attest. It is a prime example of managing customer expectation in a realistic and achievable way.

This is not 'rocket science', nor is it intensive 'Black Belt' training loaded with complex statistics. Rather, it is organised common sense. This is truly FIT Σ – 'fitness for the purpose'.

Source: Basu and Wright (2003, p. 86).

7.3 'SIGMA (Σ) FOR IMPROVEMENT' FOR SMALL AND MEDIUM ENTERPRISES

As discussed in Chapters 2 and 3, the key cornerstone of FIT Σ methodology is 'Sigma (Σ) for improvement and integration'. The 'improvement' aspect is essentially based upon the proven tools and processes of Six Sigma/Lean Sigma initiatives. As indicated in Chapter 4, we do not introduce any new 'FIT Σ tools', but rather they are more appropriately used Six Sigma and Lean tools. The improvement processes for SMEs should be proportional to the complexity of the problem and the cost–benefit trade-off of the solution. It is recommended that DMAIC Lite and Kaizen Event methodologies are adopted by SMEs in their FIT SIGMA. The learning deployment of FIT Σ for SMEs should ensure both the understanding and application of these well-developed and proven 'appropriate tools' and processes. For small and medium-sized companies, funding and skills may be an issue (Achanga *et al.*, 2006). There is also scepticism among SMEs about the benefits of Lean and Six Sigma. The SMEs appear to be more receptive to Lean than to a full-blown Six Sigma approach (ACE, 2008).

The traditional top-down implementation style, where projects are led by Black Belts, is a major barrier to entry for SMEs (Antony *et al.*, 2005). There are alternative Six Sigma deployment models that allow smaller and mid-sized organisations to implement at a faster pace than their larger customers. In a FIT SIGMA approach a Green Belt or a Yellow Belt addresses many of the constraints of SMEs and allows them to implement at a more manageable pace. As indicated earlier, some senior managers in an SME may be trained as Green Belts to lead projects.

It is recommended that DMAIC Lite methodology is applied to SMEs, as shown in Figure 7.1. The projects are aligned with the business strategy and capability of the business. Kaizen Event is also applicable to SMEs.

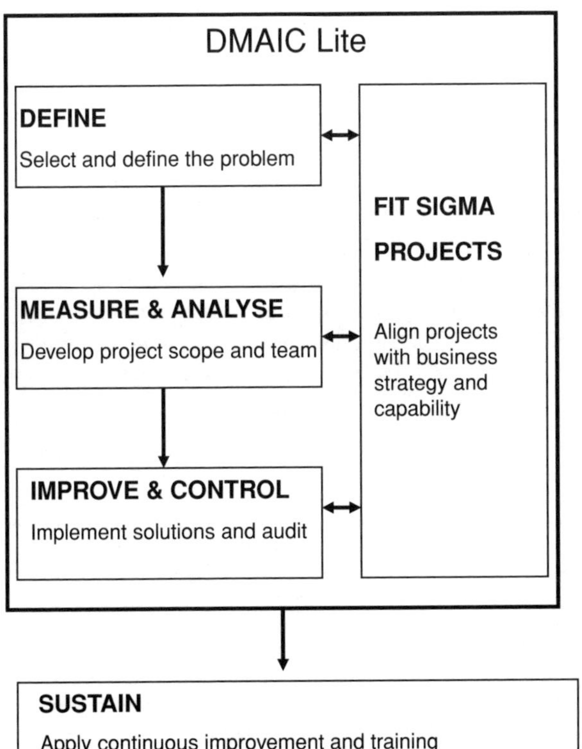

Figure 7.1 DMAIC Lite methodology for SMEs

A recommended tactic is to start with a pilot project in Lean by using a Kaizen Event approach and then gradually train in-house resources to prepare for FIT SIGMA. There are many global companies comprising small units in different countries. One such company is Rosti Technical Plastics with eight small manufacturing units in Europe and Asia. The following case example illustrates how a Kaizen Event approach was applied to Rosti's Polish facility.

Case example: Kaizen Event in Poland

Rosti Technical Plastics classifies itself as a plastic injection moulding organisation and is part of Danish AP Moller Group. The Kaizen Event (called the Kaizen Blitz) was carried out in Rosti's Polish facility in the town of Bialystok.

On the first day of the Kaizen Event the general manager announced that the goal of productivity improvement could be set at 25%. The management team at Rosti Poland found it hard to envisage any significant improvement and had to be convinced to set improvement goals above 15%. Manual kettle assembly plant was chosen. The basic process was four years old and it was widely believed that the practice could not be significantly improved.

The Kaizen team was organised with a group leader and a facilitator from Rosti Scotland as well as selected members from the Polish plant. Only a small number of the chosen Kaizen team spoke English and an interpreter was with them on a full-time basis. The Polish culture and knowledge of the plastic moulding process acted favourably upon the outcome of the event. The team members working in the process were most committed to the success and proved to be very inventive in terms of their solutions to the problems of eliminating waste from the process. The tools used were value stream mapping, SIPOC, cause and effect diagram, five Ss and brainstorming. The major root cause of the sub-assembly problem was found to be the cold embitterment of plastic parts breaking at sub-zero temperatures in Poland. These 'war room' activities lasted for five days and the new kettle assembly process was implemented within 30 days.

The final day of the Kaizen Event took place on 30 November – St Andrew's Day – and in light-hearted spirit, the team decided that the team leader and the facilitator (both were from Scotland) would wear their national dress to celebrate the success of the event. As can be seen from Table 7.2, the Kaizen Event was a success far exceeding the goals set by management and fully justified both Scots wearing their kilts!

Table 7.2

Metrics:	Goal	5 days	30 days
Productivity improvement	25%	23.5%	42%
Reduce space per kettle	45%	70%	N/A
Housekeeping (5S score)	100%	20%	280%
Reduce work in progress	50%	N/A	75%

Source: Carrick Lean Consultancy, Prestwick, Scotland.

7.4 FITNESS FOR SUSTAINABILITY

As described in Chapter 2, the fitness for sustainability of performance in FIT SIGMA is underpinned by four key processes:

- performance management;
- senior management review;
- self-assessment and certification;
- knowledge management.

The most important process for sustaining performance in an SME is knowledge management. However, this will need to be adapted to the small and medium-sized companies, taking their resources into consideration. Other procedures should also be proportional to the availability and capability of resources, as a major constraint in SMEs is the availability of resources and their appropriate skills. Achanga *et al.* (2006) point out that many smaller companies employ people with low skills. Their lack of education is probably compensated by the process of successfully training employees. Adding *people and partners* and *problem solving* from Liker's 4P model (Liker, 2004) to knowledge management could make the sustainability of results yet more effective.

The robust role of knowledge management in sustaining performance standards is illustrated by the continuous learning programme at General Cable, Mexico.

Case example: Continuous learning at General Cable, Mexico

When one walks the floor at the General Cable plant in Piedra Negras, Mexico, the tangible example of sustainable operator-led process control is visible. In the raw materials cell, employees are responsible for managing nearly 6000 active raw material SKUs. The plant's nearly perfect inventory accuracy rate over the past four years is attributed to the fact that workers are empowered to manage entire aisles in the raw materials area. Management trust in the capability of workers or associates is underpinned by a continuous learning programme.

The plant's training programme includes advanced safety instruction, education on industry standards and an extensive Six Sigma certification process. Associates are encouraged to become multifunctional and acquire Green Belt status by gaining formal certification, and 90% of employees are cross-trained in two or more operations. This knowledge management programme also requires all associates to attend an introductory Lean Manufacturing course, and it gives them several outlets to use their Lean training, including a Kaizen suggestion system and a process-improvement request scheme. Employee ideas and proposals have helped the plant achieve high marks in quality and generate nearly $1.4 million in cost savings through improvement projects in 2008. The company also recorded four million hours without a lost-time accident/incident.

Source: www.industryweek.com, January 2010.

7.5 SUMMARY

FIT SIGMA projects for SMEs should be derived from the strategic plan of the business. Based on key business goals, senior management defines the scope of each FIT SIGMA project and organises the right project team. The methodology applied should be DMAIC Lite and Kaizen Event. The training programme should also be proportional to the capability and requirement of the business and employees should be trained for Green Belt and White Belt levels. While projects are progressing, management should review them and promptly provide financial and technical support. After these projects are completed, senior management audits the results, and establishes necessary systems, especially training, to sustain the improvements.

There are also some FIT SIGMA implementation tips for SMEs. The resistance to change in an SME culture cannot be underestimated. With a good training and communication framework it is the employees who are empowered to come up with problems and find their own solutions. An external consultant can be used to start the project and provide initial training. The implementation itself should be done by internal resources.

FIT SIGMA tips for small and medium enterprises

1. Assess where you are in quality management and continuous improvement initiatives and align with business strategy and objectives.
2. Identify the Champion or Torch Bearer.
3. Carry out a value stream mapping for key products and start with Lean to eliminate waste.
4. Select consultants/trainers for Green Belt and Champions training.
5. Train a senior manager as a Green Belt to lead projects.
6. Deploy DMAIC Lite or Kaizen methodology depending on available resources.
7. Use FIT SIGMA basic tools for all projects.
8. Execute and review projects and monitor costs and benefits.
9. Develop in-house capability and seek consultancy/training support as required.
10. Sustain the initiative and culture by performance management, senior management review and continuous training.

8

FIT SIGMA in Project Management

8.1 INTRODUCTION

Interest in Six Sigma is growing rapidly within the professional project management community, and the most common question emerging from that group is, 'How does Six Sigma relate to the Project Management Body of Knowledge (PMBoK)?' PMBoK (2006) concludes that Six Sigma and PMBoK do have connections, similarities and distinctions, and it is clear that Six Sigma complements and extends professional project management, but does not replace it. Both disciplines make important contributions to successful business outcomes. As described earlier, the core methodology of Six Sigma, i.e. DMAIC (Define, Measure, Analyse, Improve and Control) is closely linked to the approach, rigour and stages of life cycle of project management.

Even today, project managers are not comfortable with embracing Six Sigma to administer their endeavours. Arguments against it include the contention that a project is unique, a one-off, and does not have a stable process. They say that Six Sigma is only effective in repetitive stable processes. Sceptics also pose the question: do we need the data-driven statistics of Six Sigma in projects where contractors are busy just doing their jobs? Our response to such doubts and misgivings is that Six Sigma can be very effective if the tools and methodology are applied appropriately (fitted to purpose). In *Quality Beyond Six Sigma* (Basu and Wright, 2003), chapter 8: 'Project Management and FIT SIGMA' addresses the issue of fitness for purpose. In projects, we have many repetitive processes and/or there are many processes requiring design. In both situations, DMAIC or DFSS can be applied. However, the caveat is the suitability and for this reason we recommend the Six Sigma methodology for larger projects with a longer duration, projects with large management organisations or multinational contractors.

DMAIC has added the rigour of project life cycle to the implementation and close-out of Six Sigma schemes. Figure 8.1 shows the relationship between DMAIC and a typical project life cycle.

Project organisations are showing a positive interest in Six Sigma, and courses and conferences are on offer for project members. Bechtel was one of the early users of Six Sigma in delivering their multinational assignments, as the following case example illustrates.

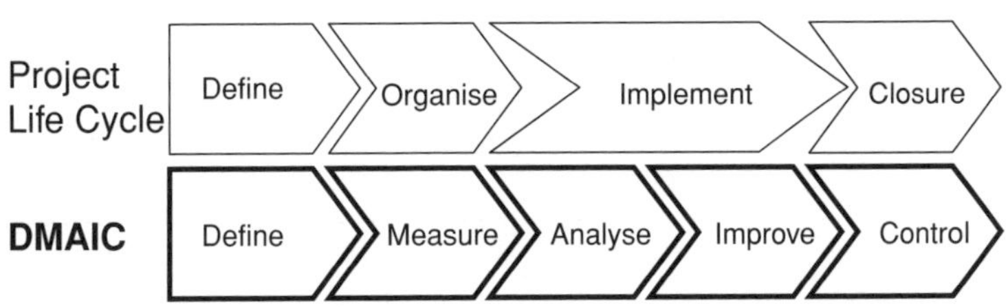

Figure 8.1 DMAIC cycle and project life cycle

Case example: Six Sigma at Bechtel

Founded in 1898, Bechtel is one of the world's premier engineering, construction and project management companies. 40,000 employees are teamed with customers, partners and suppliers on a wide range of schemes in nearly 46 countries.

Bechtel has completed more than 22,000 projects in 140 countries, including the Hoover Dam, the Channel Tunnel, Hong Kong International Airport, the reconstruction of Kuwait's oil fields after the Gulf War and the Jubail industrial city.

Bechtel was the first major engineering and construction company to adopt Six Sigma, a data-driven approach to improving efficiency and quality. Although it was originally developed for manufacturing companies, the company was confident that Six Sigma would work in professional service organisations such as Bechtel. Six Sigma has improved every aspect of Bechtel's business, from construction projects to regional offices, saving time and money for both customers and the company.

Six Sigma uses a rigorous set of statistical and analytic tools to produce dramatic improvements in their work processes, see Basu and Wright (2008). Bechtel launched Six Sigma in 2000, when the company was experiencing unprecedented growth – and facing corresponding process challenges. The company has now implemented Six Sigma in its key offices and business units around the world. About half of its employees have had Six Sigma training, and most of its major projects employ its methods from start to finish.

The investment of Bechtel in Six Sigma reached the break-even point in less than three years, and the overall savings have added substantially to the bottom line, while also benefiting customers. Some examples:

- On a big rail modernisation project in the UK, a Bechtel team used Six Sigma to minimise costly train delays caused by project work and reduced the 'break-in' period for renovated high-speed tracks.
- At a US Department of Defense site in Maryland, Six Sigma helped achieve significant cost savings by streamlining the analysis of neutralised mustard gas at a project to eliminate chemical weapons.
- To speed up the location of new cellular sites in big cities, Bechtel developed a way to let planners use computers to view video surveys of streets and buildings, making it easier to pick the best spots.
- In a mountainous region of Chile, Six Sigma led to more efficient use of equipment in a massive mine expansion, with significant cost savings.
- In the 'High Speed 1 Project', a prominent flagship project for high-speed 'Eurostar' trains between London and Brussels and Paris, Bechtel (in partnership with Arup, Halcrow and Systra) applied Six Sigma and achieved a saving of £40 million in over 500 improvement schemes.

'Six Sigma is the most important initiative for change we have ever undertaken. We're happy to report that it's becoming "the way we work".'

Source: www.bechtel.com (2006).

The application of FIT SIGMA in project management is illustrated below, with two approaches:

1. FIT SIGMA principles in addressing project management objectives (Wright and Basu, 2008).
2. FIT SIGMA in a major project (High Speed 1) (Basu, 2010).

8.2 FIT SIGMA PRINCIPLES IN ADDRESSING PROJECT MANAGEMENT OBJECTIVES

In addressing project management objectives, FIT SIGMA covers three aspects:

- What is required for a specific project to be fit.
- What has to be done to get fit.
- Once fit how to stay fit.

8.2.1 Project Management Objectives

Common problems in the management of projects are the over-run in time and budget and incomplete achievement of the original scope of the project. Ventures generally have four basic elements: scope, time, budget and quality. The well-known 'golden triangle' of project management comprises time, cost (budget) and quality. Subsidiary to these, and cutting across all four, are work break down (activities), milestones, responsibilities, cost estimation, control of costs, estimation of time, scheduling time and resource, controlling time, risk identification and management, controlling risk, balancing objectives, execution and control, finalisation and close out, follow-up after hand-over, team leadership and administration, and choice of information system (Turner, 1999). With the FIT SIGMA philosophy, a whole systems approach is taken to each of the *four basic elements* and subsidiary issues listed above.

8.2.2 Terms of Reference

The 'terms of reference' establish the overall scope, budget and time frame, i.e. the three key elements of the project. The brief follows the terms of reference and gives depth to them. The terms of reference gives what is required, while the brief identifies what has to be done to make the project happen (Turner, 1999). The requirements of the brief are reasonably accurate estimates of resources and key steps or tasks, and the skills required for each step. The brief will also endeavour to establish for each step cost, time and precedence. It is likely that the brief will also consider responsibilities and authority for the supply of resource. The brief should not be limited to the above but should include any issue that will affect the successful outcome of the project, such as establishing stakeholders. Once stakeholders are identified, especially those who are not enthusiastic concerning the outcome, the seasoned project manager will seek to find what the concerns are and if possible to reassure them or find ways around their apprehensions (Obeng, 1994; Turner, 1999).

All of the above, especially the brief, are based on estimates. By definition, each project is unique, and seldom can any planned activity be taken as a certainty. It is a recognised fact that many information technology-type projects are not completed as per the original terms of reference – indeed, many are never completed at all! A well-published example is the UK government's undertaking to introduce smart cards for social welfare beneficiaries. This project ran for several years and was finally abandoned in 2000 at a cost, according to the

National Audit Office, of a billion pounds to the British taxpayers. When the scheme was abandoned it was said that even the first three months' target had not been achieved.

The very valid reasons given by project managers for these types of failure are often:

1. the client didn't know what they really wanted, and
2. the client kept on changing their mind and adding extra features.

In reply, the customer could well say that the project manager did not listen and/or understand the customer's needs. The truth will be somewhat greyer: there will be imperfections on both sides due to imperfect communication. Irrespective of who is at fault when things go wrong, ultimately it is the reputation of the project manager that will be at stake.

For a short-term project such as building a house, no matter how carefully the plans have been drawn up to meet the client's wishes, once construction begins the client will see that an additional window would make sense, or that a door has to be moved and so on. After all, a plan is one-dimensional and reality takes a different perspective once that paper plan begins to transform into a three-dimensional product. Such changes, if agreed on early enough, will not cause problems for the builder, and should not add appreciably to the cost. For longer-term schemes such as the British government's beneficiary payment proposal, over time the sponsor will not only change requirements, but the client representatives that the project manager has been dealing with will move on to be replaced by a new group. Indeed, the beneficiary project began under the Conservative government and continued under the Labour administration. One can imagine how many changes there were in the personnel of the client, at all levels of management from ministerial level down and each with their own ambitions and bright ideas!

8.2.3 FIT SIGMA Recommendations for Achieving Scope

Firstly, it is assumed that the project has been properly constituted with the appropriate signed off terms of reference.

The first scope recommendation is to be generous in estimating the resources and time needed for inclusion in the brief, and then make sure that the client understands that, due to the novel nature of projects, estimates of time and money for resources are based on 'best guesses'. This is simply due to the fact that each project is unique and every individual venture will have its own set of unexpected problems. Some clients will press for a fixed-cost project. If the undertaking is relatively simple, for instance building a house where materials can be calculated and costed, then this might be possible. But even here allowance should be written in to enable the builder to recover any major price increases of materials. It will also allow for other contingencies, such as problems with foundations, water tables, etc. It does not serve the client well if the builder goes bankrupt and walks off the job!

The second scope recommendation is to communicate with the client. Project managers have to remember that they don't 'own' the project. They are providing a service on behalf of the client. When the terms of reference were first written it could well have been that the client emphasised finishing on time as being crucial. This does not give the project manager carte blanche authority to spend extra money above budget in trying to make up lost time when delays occur. Likewise, if it becomes apparent that the specified completion date is under threat, the project manager has a duty to advise the client as early as possible.

The third scope recommendation is to be meticulous in providing variation reports. If a client asks for a change – e.g. would it be possible to add, or amend, or whatever ... – and the project manager enthusiastically agrees, often the changes are made and the project manager believes that the client has given a firm directive to go ahead. But eventually there is a day of reckoning and the client gets the bill. This will be a problem if, when the client asked for the variation, they did not realise that there would be an extra cost involved. For example, at an early stage of house construction a request to move a window a metre to get a better view might not seem a big effort (after all, the window exists and the house is still at the framework stage of construction), but this could well take the builder several hours of labour, for which he will charge. The culmination of several such minor changes, all at the request of the client (and perhaps even some suggested by the builder) might add up to several thousand pounds extra that had not been budgeted for by the client.

The FIT SIGMA approach is that no matter how sensible the suggestion and regardless of how minor the cost, for each variation to the brief, cost variation advice must be issued to the client before that change is made.

8.2.4 Weekly Reports

With FIT SIGMA a weekly written (rather than emailed) report to the client is standard practice. The communication should be as short as possible. Two pages will be sufficient, and on distinctive coloured paper (for example bright yellow), so that the client will readily recognise (and subsequently find) each weekly report. Page one should give the actions planned for the previous week and the actual achievements of that week. Page two will show the planned activities for the next week. If there is a variance between what should have happened, and what actually happened for the previous week, a brief reason should be given as to why this discrepancy occurred, together with an outline of the effect on the overall project. With FIT SIGMA it is not sufficient merely to let the client know that the project has experienced a problem (such as one resulting in a time problem), but to recommend a solution. The proposed resolution should include answers on how to deal with time and its impact on costs.

Weekly reports benefit both parties:

1. If the client is eager for the weekly report then it will reinforce to the project manager that the client is serious and a sense of urgency will be fostered. The all-important momentum is maintained.
2. It will enable the project manager to ask early for any extra resources.
3. If problems are known and shared by both parties as they happen, or as they begin to emerge, then remedial action will be a joint decision, with no late and unwelcome shocks lying in store for the client.
4. The project manager will be encouraged to be 'up-front' about issues and will feel less alone.

Frequency and the accustomed regularity of communication will in itself help to break down barriers and create a sense of togetherness.

8.2.5 Scope Planning

Turner (1999) explains the mechanics of scope management. He emphasises that the purpose is to ensure that adequate work is performed and that unnecessary work is tackled. In FIT SIGMA parlance this means that the purpose of the project must be kept firmly in mind and for every proposed action it has to be asked: is this really necessary for the achievement of the project? This practice helps to keep things on-task.

8.2.6 Milestones

The next level of planning is the determination of milestones. This term means that such landmarks are each intermediate products or deliverables, which build to the final objective of the project. The benefits of milestone planning are:

1. To set controllable chunks of work.
2. To show how each chunk of work is related and builds towards the final objective.
3. To set fixed targets.
4. To foster a common vision for all those involved, including contractors and sub-contractors.

With the FIT SIGMA philosophy, the milestone plan is transparent for all to see.

1. For the client, it will enable them to follow progress.
2. For the project manager, it provides a means to monitor and control progress.
3. For team members, it will enable them to understand their responsibilities.
4. For everyone, it will show the overall vision.

A good milestone plan is understandable to everyone, is controllable both quantitatively and qualitatively, and focuses on necessary decisions. Turner recommends that no matter how large the project, there should be no more than 25 milestones and with no more than four result paths. Regardless of the size of the project, limiting the milestones to no more than 25 gives an easily comprehensible picture of the whole and each milestone thus becomes manageable. Every one of these signposts is made up of blocks of work. Obviously, within these portions of work there will be key areas to be monitored, and each section will in turn be broken down to have its own set of milestones (Turner, 1999).

With FIT SIGMA, detailed planning for each milestone before the project begins is NOT recommended. The approach is to only prepare fully detailed plans for activities that are about to start. Planning 12 months in advance will be out of date within a matter of weeks.

8.2.7 Time

With computer packages it is possible to calculate and show Early Start, Late Start, Baseline Start, Schedule Start, Actual Start, Duration, Float, Baseline Float, Remaining Float, Remaining Duration, Early Finish, Late Finish, Baseline Finish, Schedule Finish and Actual Finish for each activity. Baseline refers to the original plan and normally should not be changed. The other times and floats should be upgraded as each activity is completed. It is not uncommon for people to say that their project has come in on time and on budget, but ironically to overlook the fact that the baseline has long been discarded. Thus, in truth they are only coming in on time and against budget due to repeatedly changing the schedule at each review meeting!

A modification in any one of scope, time and budget areas will cause a change in one of the others (Figure 8.2). For example, more scope will mean more time and more cost and less scope might mean less time and less cost. But when things go wrong, such as a key activity has fallen behind schedule, the decision will be between reducing scope and finishing on time, or adding extra resource (cost) to try and make up lost time, or not changing the scope and finishing late. The final option of not changing the scope and finishing late will generally result in increased cost. Thus, although time in itself is not the 'be all and end all' of a project, the reality is that a time delay might mean a change in scope and almost certainly will add to the cost.

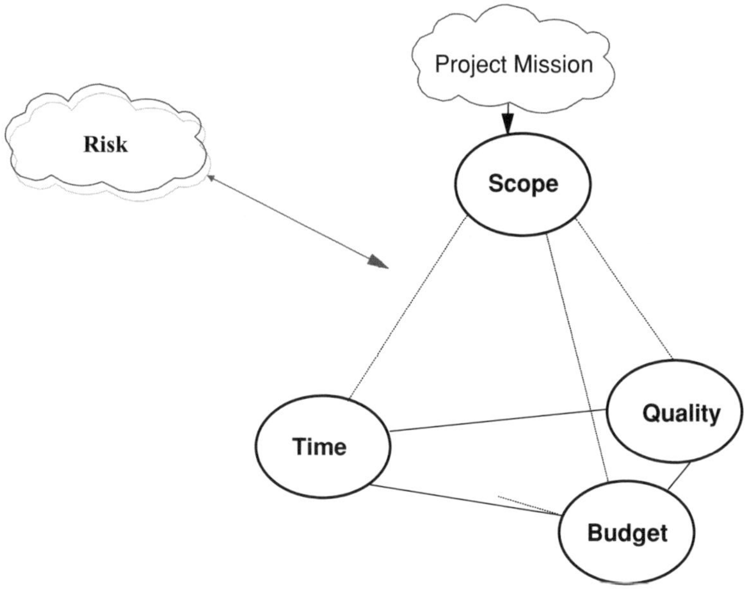

Figure 8.2 The inter-relationship of scope, time, budget and quality

The management of time begins in knowing the desired completion date, and working back and determining the point by which each milestone must be finished if the overall target date is going to be achieved. From this backward pass at scheduling, the amount of time available for each milestone and for each subordinate activity making up the chunks of work can be calculated. Knowing how much time is available for each activity will have a bearing on how much resource will be needed. As several tasks can be carried out in parallel, generally shown as 'paths' on a precedence diagram, it will be found that there is a 'float' of spare time for some activities within the overall time limit of the project. Some activities will have no float, and these actions will be critical to the overall project completing on its due date. The obvious approach is to give these critical activities special consideration so that they do not fall behind schedule and delay the completion date. The problem then arises that if other activities are not sufficiently monitored, delays can occur for these activities. Consequently, they can fall behind schedule to such an extent that they in turn put achievement of the desired completion date in jeopardy.

Obeng (1994) recommends allowing a buffer of time between every activity, including critical tasks, when scheduling resources. He says that this buffer should be considered as a separate action in its own right. Goldratt (1997) demonstrates how float can be lost when subsequent activities are not resourced so as to take advantage of an early completion right through the whole project.

With FIT SIGMA, the six steps necessary to control time are:

1. Set the start time, the amount of time (duration) and the finish time for each activity. Treat 'float' as a separate activity and where there is no float, build in a buffer activity. This will become the baseline schedule which should not be changed.

2. Do not be overly concerned with the calculations of early finish/late finish scenarios, but concentrate on the actual progress of each activity.
3. Manage and schedule resources to be available for the start date, monitor to ensure that activities finish on time, but by having a built-in buffer do not be unduly concerned if some of the safeguard for each activity is consumed.
4. At the end of each activity update the schedule, but do not change the baseline. If a buffer has not been consumed, move all activities forward and schedule resource accordingly. Add unused float or buffer to subsequent buffers (unless already behind the baseline).
5. If an activity falls behind schedule and the buffer or float is in danger of being used up for that activity, make the client aware of the situation and the likely effect on whether the final baseline is going to be achieved.
6. When delays occur that will affect the overall finish date, after consultation with the client agree on remedial action.

8.2.8 Budget

Resources cost money; thus working out the time schedule provides the basis for cost calculations. As generally there will be a budget in the terms of reference and a time line, these times and costs are mutually dependent – a change in one will almost certainly mean a change for the other. In the same manner as the time baseline was calculated, a budget baseline should also be set at the outset, which will include a budget for each activity. The control of the baseline budget is similar to the monitoring and control of any expense plan. Computer printouts will provide the budget to date, actual to date, variance for the whole project and likewise for each activity and for each milestone. The aim should be to get budget and actual cost reports as soon as possible, but often it will be several weeks before invoices have been received from suppliers and entered into the computer. Thus budget reports are received after the money has already been committed.

8.2.9 The FIT SIGMA Approach to Controlling Costs

The FIT SIGMA approach is for the project manager to identify both the big costs and the fixed costs, and to make a daily allowance for all other expenses. Wages (a major outlay for most projects) will be a known expenditure and, if keeping to the baseline, should constitute a fixed cost. The wage amount for each person used on the project should be calculated in advance to give a daily or even hourly rate. Another major expense will be sub-contractors, but again their charge on a daily basis should be known in advance. The hire of special equipment might be a major cost, but there is no reason why this cannot again be calculated on a daily basis in advance. All other charges should be allowed for as one fixed figure, which can be termed the 'ongoing cost'. This ongoing cost can be allowed for at a daily rate based on actual expenses incurred in previous projects. The ongoing cost figure should be a constant, until actual costs are reported by the accountant (usually six weeks after the event). Once these actual costs are known it might be necessary to increase the ongoing cost daily figure.

The project manager can have a 'feel' for the daily cost of a project while it is still in progress, when the daily wages and the daily fee for hired equipment and sub-contractors is identified (there is no valid reason why these expenses can't be known in advance), and if a daily allowance for all other costs is added. It might be necessary to have an assistant to keep this record, but if the project manager concentrates on only the major costs and does not get bogged down in precise details to the last decimal place, then a rough calculation can be made daily within 30 minutes.

If it is obvious that costs are ahead of the baseline budget then the client should be informed immediately. If the costs are not recoverable from the client, then the project manager's senior management will need to be informed! It is a forlorn hope that costs can be recovered at a later stage. Even if it is thought that savings can be made, it is best to be upfront when the problem occurs, and to suggest remedial action. This last point, proposing solutions to rectify the situation, is very important. It is not enough merely to say we have a problem, it is the duty of the project manager to suggest ways of alleviating that situation.

8.2.10 Quality and FIT SIGMA

Quality in projects as perceived by the client is generally based on intangibles. It will be taken for granted by the client at the outset that the scope, budget and time are met. Thus, quality from a client's perception refers to the basics of scope, cost and time *plus* the intangibles of the working relationship with the team: the ability of the project team to accommodate changes to the scope; open communication; ease of transfer/implementation from project to ongoing operation; training and follow-up service after hand-over. For tangible projects (such as a construction project), quality will also include a judgement on the standard of finish, cleanliness of the site, etc.

From the project manager's point of view, quality includes all of the above plus the costs of non-conformance resulting in delays, overtime, rework, wasted materials, idle time, putting right, etc. Quality and the culture of FIT SIGMA quality is a huge subject, beyond the scope of this chapter, but in essence it is the well-accepted philosophy of TQM. One of the key issues for project managers is the building of team spirit and fostering a quality culture with all sharing the same 'can do' philosophy. It goes without saying that in a FIT SIGMA programme the safety and health of people doing the work must be of paramount importance to the programme leader. The regulatory and company guidelines to Health, Safety and Environment must be applied to projects.

8.2.11 Hand-Over Follow-Up

If, through poor management of the client or lack of training of operators, the project does not achieve full ongoing benefits, then these shortcomings will be blamed on the project manager. It is therefore in the project manager's interest for the completed scheme to work the way it was intended. A good project manager follows up and provides 'after-sales service'. This is nothing less than sound business practice and can lead to further custom from the client, or to referred business.

8.3 FIT SIGMA IN A MAJOR PROJECT (HIGH SPEED 1)

The application of FIT SIGMA in a major project is illustrated by the case study of the High Speed 1 project (Basu, 2010). Six Sigma was applied in this venture and the FIT SIGMA approach also conforms to this application.

8.3.1 Fitness for the Purpose

The application of Six Sigma or FIT SIGMA in a major project is justifiable when the duration of a project is longer (e.g. two to five years) and a budget for Black Belt and Green Belt training is cost-effective. The undertaking in this situation may be considered as an enterprise. However, for a smaller project Lean Thinking and the Kaizen Event could be appropriate.

In 1996, the London and Continental Railway (LCR) was awarded the concession to build the Channel Tunnel Rail Link (CTLR), which was later known as the High Speed 1 (HS1) project. UR (Union Railways) was allocated the responsibility to act as the client or sponsor of the project, and RLE (Rail Link Engineering) was appointed as project manager. RLE was a consortium of construction companies formed between Arup, Bechtel, Halcrow and Systra. The favourable Six Sigma experience of Bechtel was the primary driver for the application of Six Sigma to this project.

HS1 is the first new railway in England for over 100 years. It is the first high-speed railway (capable of 300 kmph) in the UK ever, comprising 100 kilometres between London and the Channel Tunnel near Dover, with three stations and two depots. At the height of construction in 2001, the combined workforce of UR and RLE was over 1000, while at the closure stage this figure was 350. The completed £5.8 billion project (excluding the regeneration figures) opened on time and within budget. The scheme was split into two parts: Section 1 was started in 1998 and runs between the Channel Tunnel and Ebbsfleet (near Gravesend) in Kent. It completed on time and on budget in 2003. Section 2 comprised primarily the activities in the North to South network from St Pancras, and was delivered for commercial operation to Eurostar on 14 November 2007.

The original objectives of the project included:

- To provide the main high-speed railway link between Britain and continental Europe.
- To provide the transport spine and stimulus to the rejuvenation of the derelict inner city areas around Stratford and Kings Cross.

8.3.2 Sigma (Σ) for Improvement

Bechtel was the first major engineering and construction company to adopt Six Sigma, a data-driven approach to improving efficiency and quality, in major projects. On big rail modernisation schemes in the UK, including the HS1 project, Bechtel teams used Six Sigma to minimise costly train delays caused by project work, and in so doing reduced the 'break-in' period for renovated high-speed tracks.

The introduction of Six Sigma to the HS1 project delivered both cost savings and programme benefits. The Six Sigma programme trained 23 Black Belts and around 250 Green Belts and Yellow Belts. A further 100+ senior managers were educated to act as Champions in improvement projects. Over 500 such improvement projects were completed, which led to a cost saving/avoidance of at least £40 million. These schemes covered and benefited a wide range of activities across the whole HS1 project, including numerous architectural, civil and railway construction activities. Consequently, this ensured timely third-party methodology approvals, thus facilitating procurement, accelerating drawing reviews and allowing the timely generation of construction record documentation. It is evident that some of the improvement projects, such as the reduction of lead time in methodology approvals and drawing reviews, also applied Lean Thinking concepts.

UR and RLE jointly submitted qualification reports to the British Quality Foundation (BQF) for an EFQM (European Foundation of Quality Management) Award. The reports were presented in two sections – Part 1 comprising enabling factors (such as Leadership, Policy & Strategy, People, Partnership & Resources and Processes) and Part 2 detailing results factors (such as People, Customers, Society and Key Performance Indicators). The proposal was assessed externally by qualified EFQM assessors from BQF and the HS1 project received the EFQM Award in 2007. The assessment revealed that there were more strong points of the project in enabling features but weaker aspects in their results factors. The HS1 project also received the 'Project of the Decade' award in February 2008 from London Transport Award.

8.3.3 Fitness for Sustainability

As part of the HS1 audit strategy, pre-audit health checks were conducted by site teams using the same appraisal brief and checklist. These pre-audits helped the preparation of the review at each site, and also ensured that the auditees completed the outstanding Section B (for agreed corrected action) of CAR (Corrective Action Request) forms with scheduled completion dates.

In addition to these pre-audit health checks, the project team (RLE), client (Union Railways) and contractors all developed a self-assessment culture as part of their qualification for EFQM accreditation. However, there was no specially designed checklist for self-assessment, and only the EFQM criteria were applied. Following appropriate training, periodic health checks were carried out for both enabling and performance criteria of EFQM, and the HS1 project was awarded the British Quality Foundation Excellence Award in 2007. Self-assessment was used primarily to meet EFQM requirements but not to identify skill gaps for people development.

When project leaders identified the need to recruit a new member to the team, suitable candidates from member companies were sought. All positions had job descriptions detailing education, skills and experience requirements. The member companies possessed extensive people resources world-wide and a wide range of specialist skills were available when required. A survey in early 2007 showed that 23 nations were represented in the venture. According to an internal report, the project worked in excess of 82 million hours without any time lost due to industrial relation disputes.

A clear distinction was made between training that was specific to the delivery of the project (which was funded by the project) and training for personal development, and career progression which remained the responsibility of member companies. A good example of project-specific training was the comprehensive approach to health and safety training, including site inductions. Six Sigma training was given to 250 people who were also empowered to identify problems, create a business case, investigate the difficulties and instigate improvements.

Member companies of RLE had their own process for dialogue and sharing best practices. For example, Bechtel used an Employee Forum with the minutes made available on their own intranet. Arup had a similar process, known as 'Airtime'. Presentations were regularly given by leaders on the progress of the project, with a summary of successes and the challenges that lay ahead.

Even after the opening of St Pancras on 14 November 2007, HS1 maintained a skeleton project team for over six months to ensure the legacy and sustainable outcome of the scheme.

8.4 SUMMARY

It is evident that there are significant potentials and benefits to be gained from adapting and adopting operational excellence concepts such as Six Sigma (Wright and Basu, 2008; Bechtel, www.bechtel.com, accessed 14/03/08) in major projects. There are many publications showing the acceptance of supply chain management in projects (Ala-Risku and Karkkainen, 2006; Boddy and Macbeth, 2000), but applications of Lean Thinking and Six Sigma appear to be at their early stages.

Because of the perception of the one-off unique nature of a project and the repetitive nature of operations, the mindsets of many project managers may have a prejudice against – or lack of confidence to adopt – operational excellence concepts. Leading project contractors such as Bechtel, however, are also training their managers with operational excellence skills (such as those leaders tutored as 'Black Belts'). In a major project or programme with a time span of two years or more, it could be argued that the activities, processes and objectives in the organisation could be closely related to those of operations management. This could be conducive to the application of operation excellence concepts.

As we have seen, there was evidence of the excellent application of Six Sigma and EFQM in the HS1 project. Bechtel teams used Six Sigma to minimise costly train delays caused by project work and thus reduced the 'break-in' period for the renovated high-speed tracks. The introduction of Six Sigma to the HS1 project delivered both cost savings and programme benefits and conformed to the principles of FIT SIGMA.

FIT SIGMA tips for projects

1. Assess if the project or programme duration is over two years and if so, then align with project strategy and objectives.
2. Perform a preliminary stakeholder analysis and identify the Champion or Torch Bearer.
3. Carry out a value stream mapping for key work packages in Work Breakdown Structure and start with Lean to eliminate waste.
4. Start with the project team (work package) with the most interest and influence in the FIT SIGMA programme.
5. Select consultants/trainers for Black Belt, Green Belt and Champions training.
6. Deploy the DMAIC Full methodology.
7. Use FIT SIGMA tools for all projects and subsequently advanced statistical techniques only for specific complex and critical problems.
8. Execute and review projects and monitor costs and benefits.
9. Develop Master Black Belts and in-house capability and roll out to all project teams including major contractors.
10. Sustain the initiative and culture by performance management, gateway review, self-assessment (e.g. EFQM) and knowledge management.

9

FIT SIGMA in Green Thinking

9.1 INTRODUCTION

Miss Piggy once said, 'It's not easy to be lean'. Her friend Kermit the Frog might have also said, 'It's not easy being green, either'. Organisations are facing increasing challenges to balance business performance with environmental issues. Such challenges have created a new area of Green supply chain management and Green thinking. The term 'Green thinking' refers to the way in which every aspect of an organisation – innovations, policies, operations and processes – may be considered in the context of the sustainable environment.

Environmental regulations are also changing the way in which supply chains are designed and managed. The problem is that the sheer number of rules, other influences such as changing consumer sentiment and the complexity of global trade, make it difficult for companies to decide exactly how they should respond to these pressures. Firms must also address the ethical requirements of their corporate social responsibility (CSR).

On a global scale, industrial pollution is the main contributor to the so-called 'greenhouse' effect and global warming. The greenhouse gases include carbon dioxide, methane CFCs (chlorofluorocarbons) and nitrous oxide. Some scientists believe that increased emission of greenhouse gases, particularly carbon dioxide, is causing energy to be trapped, thus raising the global temperature. Methane and carbon dioxide emissions are linked to much larger economic infrastructures and are much more difficult to regulate. In 1997 the Kyoto Treaty was drawn up in Kyoto, Japan to implement the United Nations Framework Convention for Climate Change. It largely binds industrial nations to reduce the emission of greenhouse gases by an average of 5.2% below their 1992 levels over the next decade.

It is important to note that environment and safety are not just social or political issues; they are vital ingredients contributing to the performance of an organisation. In manufacturing industries, there is much scope for environment and safety. Apart from humanitarian reasons, it is a truism that accidents cost money. Likewise many businesses and organisations are facing declining reserves of natural resources, increased waste disposal costs, keener interest in their human rights records and tighter legislation.

These rising environmental pressures and social expectations can be turned to commercial advantage if a strategic approach is taken to develop a Green thinking policy. The strategic approach of Green thinking entails complex longer-term considerations involving not just industry but environment protection as an important international issue. Industrialised countries including the USA are spending between 0.5% and 1.5% of their gross national product (GNP) on the control of pollution. Therefore, whatever the scientific evidence, or lack of evidence, it is inevitable that no organisation can in the long term hope to avoid legislation and regulations designed to support the spirit of Green thinking. In fact, well-designed Green thinking strategies can lead to sustainable business advantage.

Such a policy should be supported by a process to implement its objectives. It is an enormous subject and any attempt to make a comprehensive analysis of all the issues involved is beyond the scope of this chapter. However, quality can play its part and contribute its own specific role in this global challenge and opportunity. In the domain of quality, the specific methodology of FIT SIGMA (which as we know is evolved from Six Sigma methodology) could be such a process to support a Green thinking strategy. The aim of this study is to review some of the critical issues and initiatives of Green thinking under the following headings:

- What is Green thinking?
- Why FIT SIGMA is relevant to Green thinking.
- The roles of stakeholders.
- How FIT SIGMA can help Green thinking.

9.2 WHAT IS GREEN THINKING?

The concept of Green thinking is not new. Pearce (1992) analysed the relationship with Nature and the environmental strategies identified with the 'Green movement'. Elkington (1994) coined the phrase 'triple bottom line' as a new concept of accounting practice. The triple bottom line, or TBL (also known as 'people, planet, profit') captures an expanded spectrum of values and criteria for measuring organisational success: economic, environmental and social. TBL is not without its critics, and is now more aligned with CSR. Sustainability, or sustainable development, was first defined by the Brundtland Commission of the United Nations (1987) as 'the needs of the present without compromising the ability of future generations to meet their own needs'.

With the above background, the author has developed the 3Es model (Efficiency, Environment and Ethics) as the three dimensions of Green thinking. As shown in Figure 9.1, each

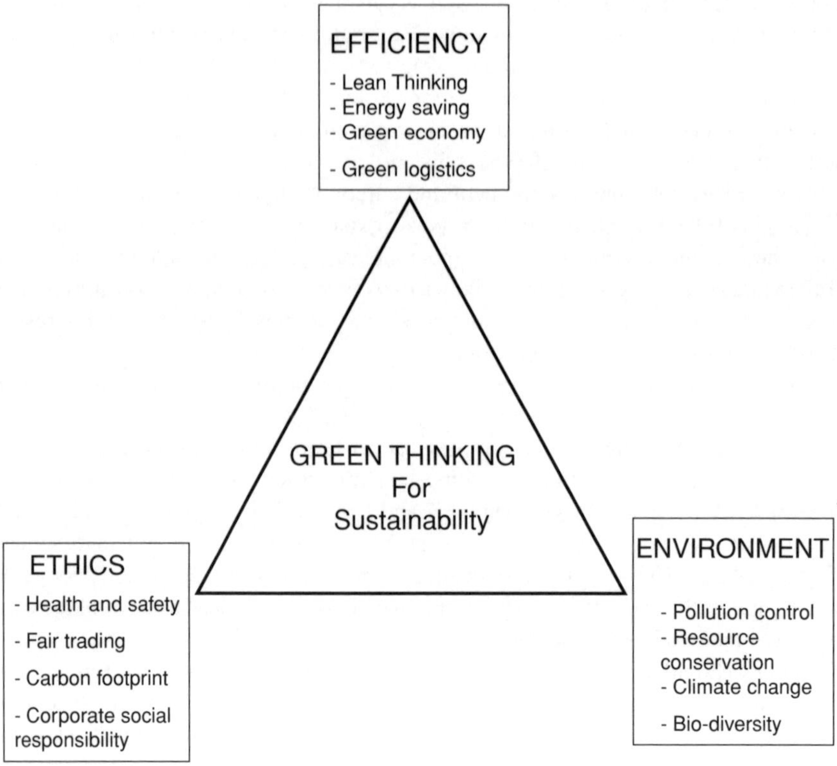

Figure 9.1 Green thinking concept

aspect includes specific attributes, e.g. efficiency relates to Lean thinking, energy saving, the Green economy and Green logistics. Better training, greater awareness and improved communication on Green issues characterise the Green economy of an organisation. Green logistics refer to buyer companies requiring a certain level of environmental responsibility in the core business practices of their suppliers and vendors. Ethics include issues and processes related to Health and Safety, fair trading, carbon footprint and CSR. The most important dimension is Environment, containing pollution control, resources (including energy), conservation, climate change and biodiversity. Climate change relates to the factors (such as CO_2 emission) causing the longer-term alterations in weather statistics. Major energy companies like BP (www.BP/sustainability) are focusing on new energy technology and low carbon energy products in alternative sources, such as wind, solar, biofuels and carbon capture and storage (CCS). The variety of life on Earth – or its biological diversity – is commonly referred to as 'biodiversity', where each species, no matter how small, has an important role to play.

The primary focus of Green thinking is on the environment and its sustainability, while efficiency and ethics contribute to the quality and sustainability of the environment. Therefore, the three dimensions and their attributes are inter-related. For example, energy saving and Green logistics in efficiency are closely linked to environment. Likewise, the pillars of fair trading and carbon footprint within ethics are intimately related to environment.

9.3 WHY FIT SIGMA IS RELEVANT TO GREEN THINKING

As discussed in Chapters 1 and 2, the significant bottom line results and extensive training deployment of Six Sigma and Lean Sigma must be sustained with additional features for securing the long-term competitive advantage of a company. If Lean Sigma provides agility and efficiency, then measures must be in place to develop a sustainable fitness. The process to do just that is FIT SIGMA[1] (Basu, 2009). In addition, the control of variation from the mean in the Six Sigma process (σ) is transformed to company-wide integration via FIT SIGMA methodology (Σ). FIT SIGMA is therefore synonymous with 'FIT Σ'. Furthermore, the philosophy of FIT Σ should ensure that it is indeed fit for all organisations – whether large or small, manufacturing or service.

Four additional features are embedded in the Six Sigma philosophy to create sustainability in FIT SIGMA. These are:

- A formal senior management review process at regular intervals, similar to the Sales & Operational Planning procedure.
- Periodic self-assessment with a structured checklist which is formalised by a certification or prize, similar to the EFQM Award but with more emphasis on self-assessment.
- A continuous learning and knowledge management scheme.
- The extension of the programme across the whole business to ensure Green thinking, with a shift of the theme from the variation control (σ) of Six Sigma to the integration of a seamless organisation (Σ).

Green thinking can leverage the contents of FIT SIGMA to achieve specific environmental and business goals. For example, the value stream mapping of FIT SIGMA (to pinpoint waste) can be adapted to create Green value stream mapping (to identify carbon footprints). Likewise, 'voice of environment' can represent the growing prevalence of environmental drivers not captured in 'voice of customer'.

[1] FIT SIGMA is a registered trademark of Performance Excellence Ltd.

As a comparative note, 'Green Sigma', developed by IBM, employs the Lean Six Sigma principles wherever energy and water are used throughout a client's operations (*Quality Digest*, 15 September 2008).

9.4 THE ROLES OF STAKEHOLDERS

The key stakeholders involved in Green thinking are manufacturers and suppliers, governments and non-profit organisations, retailers and consumers. Their roles are reflected in respective initiatives as summarised below.

9.4.1 Green Initiatives by Manufacturers and Suppliers

It is reasonable to state that manufacturing industries are major players regarding environment topics. But when the issues relate to safety, whether for products or workplaces, they apply seriously to both manufacturing and service organisations. Lack of safety in the product or in the workplace will inevitably cost money. Accidents mean lost production time plus time-wasting inspections by government officials. They may also incur legal costs, as well as the expense of correcting the situation. It has to be cheaper to do it right first time. One of the better-known environmental standards is put forth by the International Organisation for Standardisation (ISO), known as ISO 14001.

Basu and Wright (2003) have established that environmental protection relates to pollution control in two stages. Conventional restraints, or 'first generation pollution' controls, are applied to pollution in air, water and regarding noise created in the manufacturing process. Such jurisdictions are usually regulated by legislation. There is also a 'second generation pollution', which relates to the problems caused by the usage of certain products and chemicals over a long period. The most widespread example of such 'second generation pollution' is the contamination of land which permeates ground water.

9.4.2 Green Initiatives by Governments and Non-profit Organisations

The UK government has set a target for the central government office estate to achieve carbon neutrality by 2012. The UK has an overarching target to reduce greenhouse gases by 26% or more by 2020 and by at least 60% by 2050. Environmental protection is moving in cycles without showing a continuous improvement. In the 1970s the environment was a political hot potato; however, since then we have become accustomed to the prominence of Green issues. Without doubt some concerns were overstated: for example, it was widely said in the 1970s that oil would run out by 2000, and we all remember the nonsense that the YK2000 question turned out to be ... However, since the Kyoto Treaty of 1997, influenced by pressure groups such as Friends of the Earth, and the well-publicised activities of Greenpeace, environmental issues are again at the forefront. The scientific evidence of global warming produced by experts (e.g. of the Royal Society) has seriously created a sense of urgency in governments, including the State governments of the USA as well as non-profit organisations world-wide.

One such non-profit organisation in the UK is the National Centre for Business and Sustainability (NCBS), which is committed to advancing sustainable policy solutions, and has already shown the way forward through a number of ground-breaking studies. The NCBS is working with a range of businesses and organisations to help integrate the principles of sustainable development into policies, programmes and decision-making processes. The Centre does this by taking a practical approach to sustainability, combining the inspiration of The Co-operative Bank's ethical and ecological policies with a number of practical and applied tools that help put sustainability into a working business context.

The UK government has set up a state-funded non-profit organisation called Envirowise (www.envirowise.gov.uk). Envirowise delivers a valuable government-subsidised programme of free and confidential advice to UK businesses. This assistance enables companies to increase profitability and reduce environmental impact.

In addition, many countries and regions are already taking action on the environment. Both countries within the EU, and non-members of Kyoto such as the USA, Australia, China and India, have said that they will reduce greenhouse gas emissions. The UN Framework Convention on Climate Change and the Kyoto Protocol provided a basis for international co-operation on climate change initiatives. Countries are facing a range of diverse circumstances and thus will use different approaches to make their own contribution to tackling climate change. These strategies will have both direct and indirect impact on supply chain management. Key elements of future international frameworks could include emissions trading, technology co-operation, action to reduce deforestation and the adaptation of new, cleaner technologies in developing countries.

9.4.3 Green Initiatives by Retailers

Global retail giants Wal-Mart and Carrefour as well as other supermarkets all over the world are responding to the pressures on packaging waste reduction and the additional environmental issues involved in the Green supply chain.

Recent media reports are loaded with announcements on 'greening the supply chain' from large retail groups. Wal-Mart, a US company and the world's largest retailer, unveiled its packaging scorecard to major suppliers such as Proctor & Gamble, Unilever and Nestle, which was designed to cut packaging. Wal-Mart hopes that the scheme will reduce packaging across its global supply chain by 5% by 2013.

British supermarkets have taken the initiative in becoming more environmentally aware. ASDA supermarket, a subsidiary of Wal-Mart in the UK, claimed, as an example, that taking pizzas out of cardboard boxes saved 747 tonnes of cardboard in a year. UK supermarket Sainsbury's announced in October 2006 that 500 of its own-brand goods would be presented in compostable packaging. 'Friends of the Earth', a non-profit organisation in the UK, gave a cautious welcome to Tesco's new environment fund of $100 million but said the supermarket giant still had a very long way to go if it was serious about greening its operations. Tesco would need to address a number of key areas if it was serious about reducing its environmental impacts. These include moving away from car-dependent stores, switching from its global supply chain, radically improving energy efficiency in its stores and cleaning up its supply chains.

Perhaps the best example of a company devoting itself to acting upon environmental issues is the British retailer Marks and Spencer. Their website (marksandspencer.com/about) details the evolution of their 'Plan A' scheme, launched in January 2007, listing 100 commitments that the firm aimed to achieve within a five-year time frame. Since then, M&S have extended this commitment to encompass 180 goals that they wish to attain by 2015. Their proudly stated aspirations include 'working with our customers and our suppliers to combat climate change, reduce waste, use sustainable raw materials, trade ethically, and help our customers to lead healthier lifestyles', and that they have 'the ultimate goal of becoming the world's most sustainable major retailer'.

Even the airlines, the biggest polluters with their CO_2 emissions, have joined the Green bandwagon. Richard Branson committed the next 10 years' profits for Virgin – around $3bn – to fighting global warming.

There has been stronger emphasis to introduce ethical, fair-trade, organic and bio products to the consumer. The following case example of 'Carrefour Bio coffee' illustrates that, by promoting unbranded 500g/1kg coffee in bags as 'organic coffee to support fair trading' in 1997, sales increased by 80% in four years.

Case example: Carrefour Bio coffee

Carrefour is a global hypermarket retail chain organisation from France with a turnover of more than €100 billion, making it second only to Wal-Mart, the largest retail company in the world.

The first shipment of coffee beans was delivered in 10kg sacks to Vitrolles, France, in 1970. The beans were roasted in store and sold in 500g and 1kg bags. In April 1997, Carrefour launched an 'organic' coffee brand under the name 'Carrefour Bio' to promote organic products and support fair trading. In 2001 it was decided to establish a 'Green supply chain' for 'Carrefour Bio'.

The organic coffee marketed under the name 'Carrefour Bio' is not indexed on the world coffee market. The purchase price is approximately 30% higher than the average cost in Mexico. The supplier is contracted to pay a guaranteed minimum price to producers. Thus, producers can obtain up to 60% of the value of the coffee at current international rates. Some 3000 producers from 37 Mexican communities cultivate coffee using organic methods. A local infrastructure has been introduced to transport people between towns and villages (a two-hour bus ride replacing what was previously a two-day walk). A health scheme has been introduced providing free medicine and healthcare and a consortium has been set up to buy basic foodstuffs at cost price.

The coffee is cultivated by small farmers working for the Uciri cooperative in Mexico using an organic method of farming. Such cultivation helps prevent the land becoming impoverished. The cultivation is carried out in accordance with French Organic Society standards without the use of organofluoridated fertilisers or chemical pesticides for tropical forest conditions. An organic fertiliser comprising sun-dried hand-picked stoned cherries and animal waste is spread over the plants. This is the only plant treatment used by the farmers.

Cultivation methods are monitored by an organic certification body. An agricultural education centre has also been established catering for organic farming, animal breeding and bio culture.

The 'Carrefour Bio' coffee project appears to constitute a win/win initiative for the Green supply chain. For Carrefour, sales of the product increased from 29.5 tonnes in 1997 to 54 tonnes in 2001. The fertility of the land has been protected. The average income per family of producers increased from €53 per year in 1985 to €1524 per year in 2000. The local communities benefitted from the infrastructure and facilities for transport, healthcare and education. Finally, consumers are happy with an organic product at an affordable price.

Source: Carrefour Belgium (2002).

Such is the importance of a company displaying its eco credentials that there is a growing awareness that communicating their efforts to the consumer leads to favourable publicity and an enhanced reputation of the brand. Indeed, close perusal of any paper media advertising will reveal how many companies actually seem to be using newspapers and magazines to broadcast their environmentally conscious efforts instead of boasting about their products. In today's climate, ecological issues are such a current and hot topic that firms realise they can tap into the public consciousness and portray themselves as 'Green', thus superseding the traditional tenets of advertising, the focus being on the goods themselves. This method is used not only by supermarkets but also by manufacturers of white goods or motor cars (emphasis on energy efficiency of products), petroleum firms, energy companies and so on. The reader can also examine everyday objects to see this principle in action: the covers of recently published books (which inform us of their use of paper from sustainable forests), or the packaging of almost any food or cosmetic item to see details of the sustainability of their source materials.

9.4.4 Green Initiatives by Consumers

Consumers have both the power and the responsibility to enhance the activities and effectiveness of the Green supply chain. It is the consumer who pays for the end product or service, and it is the consumer who ultimately suffers or benefits from the resultant impact on the environment. Green initiatives from consumers could be manifested in three ways:

- make your home Green;
- feedback to retailers;
- reverse supply chain.

'Make your home Green' is becoming a conscious target of many consumers. This is effected in two paths. First, consumers are attempting to minimise 'carbon emission' by making houses and household appliances more energy-efficient and also by moving towards eco-friendly transport. Second, encouraged by local authorities, consumers are making commendable efforts in the recycling of household wastes by the provision of Green (compostable) waste, paper and glass bins for the home; use of local recycling centres, etc.

A reverse supply chain is a process of getting goods from the customers back to the manufacturers. It's a relatively new trend in supply chain management that focuses on 'Green manufacturing' to target recycling, recovery and remanufacturing systems. In these reverse networks, consumers bring products to a retailer or a collection centre. For example, supermarkets in Germany have a bin where customers leave used batteries. Depending on the particular product, it can be refurbished, remanufactured or recycled, making sure the physical flow is efficient. It is estimated that 63 million personal computers world-wide became obsolete during 2003 and about 10 million electric waste products are dumped per year in Japan. By 2030, the obsolete PCs from developing regions will reach 400–700 million units, far more than from developed regions (at 200–300 million units) (Yu *et al.*, 2020).

Mobile phones can be returned to the store where the new one is purchased. From there, the phones are resold and reused in other countries where the very technology that is being phased out in developed countries is being introduced. Many other products have the potential for second use, including computers, auto parts, printer cartridges, refillable containers and a host of other possibilities. In remanufacturing, reverse logistics introduces additional challenges to planning for a closed-loop supply chain. The planning, sourcing, making and delivery of the products are affected by the reverse flow of used products and materials for subsequent consumption in the manufacturing of new products. Reverse logistics play a key role as retail organisations tend to look at their reverse supply chains more closely to enhance customer satisfaction, cost/time efficiencies and supplier performance.

9.5 HOW FIT SIGMA CAN HELP GREEN THINKING

The tools, techniques and processes of FIT SIGMA can provide an effective framework to implement the initiatives in all three dimensions of Green thinking, viz. Efficiency, Environment and Ethics.

9.5.1 FIT SIGMA in Efficiency

The fundamentals of Lean thinking are embedded in FIT SIGMA and these are underpinned by:

1. Elimination of waste.
2. Smooth operation flow.
3. High level of efficiency.
4. Quality assurance.

The Lean methodology as laid out by Womack and Jones (1998) is sharply focused on the identification and elimination of 'mudas' or waste; indeed, their first two principles (value and value stream) are centred around the eradication of waste. One important area of waste in processes is excess inventory. The cycle time, or lead time, reduction is another target area of waste reduction. Environmental misuses – including squandered energy – can cost companies thousands of dollars a year. FIT SIGMA tools (such as value stream mapping and the flow diagram) are there to analyse processes and identify wastes and non-value-added activities. The additional impact of FIT SIGMA on efficiency is ensuring the sustainability of that efficiency by quality assurance and well-structured training programmes for Black Belts and Green Belts.

Case example: GE jet engine testing

A facility of General Electrics in Peebles, Ohio (where GE runs jet engine tests in the open air) is a leader within the company in applying Lean to address greenhouse gas (GHG) emissions. The quantity of testing conducted is directly related to the amount of business the airlines are doing. More engine testing means more jet fuel consumption and more GHG emissions. These reductions have also resulted in significant cost savings.

Before Lean events, the engine had to be turned on three times in order to complete the balancing process. Lean methods helped Peebles develop a new balancing process that only required the engine to be turned on once, which reduced both fuel consumption and GHG emissions. While the facility's old troubleshooting process required the engine to be running, the new procedure allowed troubleshooting to occur while the engine was not running. Through these changes the company reduced its use of fuel from 20,000 gallons to 10,000 gallons. There was also an overall cost savings of over a million dollars a year.

Source: US Environment Protection Agency (www.epa.gov).

Table 9.1 Toyota's 5R for Green thinking

5R	Measures	Responsibility
Refine	Expansion to reduce, reuse and recycle by changing design and raw materials	Operations departments
Reduce	Reduction of amount of waste generated at source	Operations departments
Reuse	Reuse within operations processes	Operations departments
Recycle	In-plant and outside use of generated waste	Operations departments
Retrieve energy	Recovery of energy from waste materials that cannot be refined, reduced, reused or recycled	Environment Technology department

The elimination of waste through Lean thinking in any entity or form of consumption, whether products, processes, materials or utilities, is de facto in the domain of Green thinking. Lean thinking is also a key component of FIT SIGMA, and thus FIT SIGMA can be a major driver of the 'Efficiency' leg of Green thinking. In the well-known Toyota Production System (TPS), the forerunner of Lean thinking, Toyota has developed a 5R programme (see Table 9.1) to reduce polluting wastes in order to maintain production with Green thinking (Black and Phillips, 2010).

It is evident from the 5R programme that eliminating waste at the source by using Lean and related FIT SIGMA techniques helps the process of achieving zero waste as part of Green thinking in efficiency.

9.5.2 FIT SIGMA in Environment

The impact of FIT SIGMA in the environment is achieved by identifying and prioritising environmentally related projects as part of Black Belt/Green Belt training programmes. There are five areas under which environmental issues are addressed in identifying FIT SIGMA projects, namely:

1. Green procurement.
2. Energy conservation.
3. Resource conversation.
4. Elimination of hazardous waste.
5. Risk management.

Green procurement is the selection of products and services that minimise environmental impacts. The rationalisation of the supply chain to significantly reduce delivery vehicle movements is such an example. Green procurement is part of Green logistics of efficiency, but its focus on environment is more significant in FIT SIGMA projects. Energy conservations are achieved through efficient energy use – in other words, energy use is reduced while achieving a similar outcome. In addition to industrial facilities, domestic housing and road transport are also two major areas of concern. The Resource Conservation Challenge (RCC) is a national effort to protect natural resources (including energy) by managing materials more efficiently. 3M is a pioneer in the use of Lean Six Sigma methods and tools to minimise hazardous waste and improve operations and quality. The famous 3P programme (Pollution Prevention Pays) of the 3M company brought about major savings, including $2 million from the elimination of hydrocarbon wastes from a reactive casting process.

Case example: 3M company – Environmental, Health and Safety

When 3M instigated its 3P programme several years ago, the approach was to capture and control pollutions and emissions before they could damage the environment. This tactic, although effective, has been altered to a philosophy of prevention rather than containment. The 3P programme now aims to prevent pollution at source by using different materials, changing the process, redesigning the plant and equipment, and recycling waste.

Since 2001, 3M's Environmental, Health and Safety operations (EHS) organisation has deployed Lean Six Sigma to improve corporate EHS services and activities. EHS team members receive two weeks of Lean Six Sigma Green Belt training, and coaching is provided by Black Belts.

There may be less obvious tools and processes of FIT SIGMA to contribute to the debate of global warning and climate change. However, that does not imply that FIT SIGMA denies the effect of human living styles in global warming. In fact, the rigour of Six Sigma and FIT SIGMA can be applied to verify the issues of the broader climate change debate. There are many FIT SIGMA tools in the book that are highly relevant. For example, control charts can help analyse the range of variability in climate change while cause and effect diagrams can articulate the different causes of global warming. At a more advanced level of analysis, Six Sigma tools like DOE (Design of Experiments), MSA (Measurement Systems Analysis) and SPC (Statistical Process Control) can contribute a data-driven rigour to the debate of climate change. As a specific example, SPC analysis of global temperature pinpointed special causes of variation affecting the temperature data (www.sigmatraining.org/tag/temperature.trends). In this example, three special causes are identified and analysed as the 'heat island' effect (urbanisation), CO_2 emission due to human activity and the sunspot cycle.

FIT SIGMA is not the 'holy grail' to environmental challenges or the climate change debate, but it can contribute as a systematic and proven methodology in these challenges. There are many other initiatives from all stakeholders that are required, as discussed earlier. There are many such initiatives in progress, such as the Renewable Energy Directive (EU target to produce 22% of electricity from renewable sources), the Climate Change Levy (tax on greenhouse gases) and the Carbon Reduction Commitment (UK target to cut carbon emissions by 26% by 2020). There are more legislations in the pipeline, including the regulation of waste in portable batteries and widening the scope of the Hazardous Waste Regulation. It is a good start, but has a long way to go.

9.5.3 FIT SIGMA in Ethics

There are three key aspects of ethics which could be supported and enhanced by FIT SIGMA, namely:

1. Product and occupational safety.
2. Carbon footprint.
3. Corporate social responsibility.

A major incident in product safety, particularly in consumer and food products, can seriously damage the brand and business. Accidents in the workplace mean lost production time and are often accompanied by paying out both compensation to employees and legal costs. A carbon footprint is 'the total set of GHG (greenhouse gas) emissions caused directly and indirectly by an individual, organisation, event or product' (UK Carbon Trust, 2008). An individual organisation's carbon footprint is measured by undertaking a GHG emissions assessment. CSR is a self-regulating mechanism whereby organisations would monitor and ensure their adherence to the law, public interests and ethical standards. CSR also embraces the concept of the 'triple bottom line', covering environmental, social and economic bottom lines. A key role that FIT SIGMA can play in ethics is bringing together the specialists in HSE (health, safety and environment) and other operations in Black Belt/Green Belt training.

Adverse publicity surrounding ethics is not an attractive proposition to any organisation, nor is its potential financial cost. In addition to implementing environment management systems such as ISO 1401 or EMAS (Eco-Management and Audit Scheme), the adherence to a strong ethical culture by continuous training means that the executive board of an organisation can be confident that brands or the business would not be tainted by prosecution or negative press. The training on environment, safety and ethics can be embedded in FIT SIGMA training workshops.

9.6 GREEN SIGMA

The rising energy costs, increasing regulatory pressures and emphasis on corporate ethics are forcing organisations to focus on the impact of Green issues on their business results. Consequently, the market for Green initiatives is growing rapidly and business customers are becoming more and more aware of Green issues. FIT SIGMA can be a method of supporting these activities, but some consulting companies are also working on products dedicated to Green issues. One such initiative is the launch of Green Sigma by IBM.

IBM announced a trademarked Green Sigma service based on the Lean Sigma process improvement methodology and designed to help firms cut the carbon intensity of their business processes. As shown below, the key steps of IBM's Green Sigma conform broadly to the FIT SIGMA approach to Green thinking.

The Green Sigma™ service of IBM

The IBM Green Sigma service consists of a five-phase approach to a client's operations and environmental practices:

1. Establish key performance indicators defined and tailored to the client's operational environment, industry and business.
2. Identify areas where activities and practices should be measured.
3. Use a Carbon and Water Management Dashboard system to monitor key performance indicators and analyse performance data.
4. Apply Green Sigma statistical tools and techniques to analyse and improve processes to reduce energy usage, carbon emissions and water inefficiencies.
5. Create ongoing optimisation of processes and key performance indicators through the Carbon and Water Management Dashboard, and identify new areas where improvements can be made.

Source: IBM Media Relations (www.IBM.com).

9.7 SUMMARY

'The scientific evidence is now overwhelming: climate change is a serious global threat, and it demands an urgent global response', concludes Nicholas Stern (2006).

It should be noted that this view has been disputed by other scientists. However, irrespective of what we believe, the pressure is on for industry and nations to adopt a Green approach to the supply chain. In this chapter we have attempted to present a balanced view of various initiatives adopted by manufacturers and suppliers, government and non-government organisations (NGOs), retailers and also consumers. Every stakeholder has a role and responsibility in 'greening' the supply chain. We have shown that there are commercial benefits in reducing wastes (e.g. excessive packaging). Large retailers like Wal-Mart, Carrefour and Tesco are probably facing disproportionate demands from environmental pressure groups and regulatory bodies, but nonetheless are displaying visible efforts to respond to these demands.

Quality as a whole can play a major role in raising the awareness of Green thinking and sustainable developments. In this chapter, only the impact of FIT SIGMA and Six Sigma programmes (which are part of the quality movement) are touched upon. Learning from the proactive focus on Green thinking in the Six Sigma programmes of leading organisations like GE and 3M, the adaptive tools and processes of FIT SIGMA could be applied appropriately by all organisations in their Green thinking initiatives. Edward de Bono (www.indigobusiness.co.uk) also suggests the inclusion of 'Green Hat' in his concept of 'Six Thinking Hats of Six Sigma'. In addition to following the fundamental tenets of Six Sigma and the additional features of FIT SIGMA in Green thinking initiatives, all organisations should be bolstered by adopting the following principles.

FIT SIGMA tips for Green thinking

1. Include specialists from HSE (health, safety and environment) and CSR (corporate social responsibility) in Black Belt/Green Belt training programmes.
2. Proactively seek improvement projects covering the three dimensions (e.g. Efficiency, Environment and Ethics) of Green thinking.
3. Apply FIT SIGMA tools and techniques (e.g. value stream mapping, flow diagrams and brainstorming) to analyse and identify non-value-added activities in projects related to Green thinking.
4. Consider order of magnitude estimates of the 'carbon footprint' of all projects in evaluating any future investments resulting from a project.
5. Aim to incorporate 'Green' aspects in conventional cost-savings projects to identify both environmental and ethical benefits as additional selling points.

10 Implementation of FIT SIGMA: Making it Happen

10.1 INTRODUCTION

When we began to write the chapter on implementation in *Quality Beyond Six Sigma* (Basu and Wright, 2003), I had just returned from a Six Sigma conference at the Café Royal London. A female delegate at the conference commented, 'A Six Sigma programme is like having a baby – very easy to conceive but difficult to deliver!' The implementation of Six Sigma, FIT SIGMA and for that matter the instigation of any change programme is indeed a bit like having a baby. It is very pleasant to conceive, but the delivery of change can be a tricky process. According to Carnall (1999), 'the route to such changes lies in the behaviour: put some people in new settings within which they have to behave differently and, if properly trained, supported and rewarded their behaviour will change. If successful this will lead to mindset change and ultimately will impact on the culture of the organisation'.

10.2 SELECTION OF TOOLS AND TECHNIQUES

Regardless of the quality programme that an organisation may choose to adopt, a selection of tools and techniques will be essential to progress the initiative. Therefore, the big question has to be: which tools to use and which ones do you start with? Here are three simple tips.

First, you should have a complete toolbox at hand. Just as a good golfer will not compete in a championship with just one iron, we recommend that you should acquire a full bag of irons (tools) and clubs (techniques). That is, you should know how to get access to the whole set. These tools and techniques are not new and they should be brought together to provide a well-stocked toolbox appropriate for the purpose. Chapters 3–9 should assist you in doing just that.

Second, to continue the golfing analogy... we should acknowledge that owning the best golf clubs does not make someone a Tiger Woods, nor does borrowing a racquet from Roger Federer make us a Wimbledon champion. In other words, success depends on the skills of the players – developed by rigorous training and practice. Therefore, it goes without saying that the employees from the top down to the bottom of an organisation, especially the key players, should be provided with the right level of education and training. This will ensure that their knowledge and understanding of the tools and techniques are appropriate for the specific process and application. We have provided general guidelines on training requirements for each tool and technique in the preceding chapters.

Third, start with simple tools first. The experience of organisations who have successfully applied Six Sigma, such as Allied Signal, Bombardier, General Electric and Dow Chemical, to name a few, showed that 80% of their projects used 'basic tools'. The success of TQM in Japan also indicated, according to Ishikawa (1985), that 'simple' methods led to 90–95% of process improvements. So what are these simple tools that we should begin with? We suggest a 'starter pack' which could be used in training workshops for 'Leadership', 'Awareness' and 'Green Belt'. Chapter 4 suggests the following set of appropriate 'FIT SIGMA tools':

value stream mapping
SIPOC diagram
flow diagram
CTQ tree
project charter
run chart
histogram
cause and effect diagram
Pareto chart
control charts
regression analysis force field analysis SWOT analysis
PESTLE analysis
five whys
inter-relationship diagram
SMED
five Ss
mistake-proofing
brainstorming
OEE
Gantt chart
activity network diagram
radar chart
PDCA cycle

Only the control chart from the above list would require a good understanding of statistical process control. Depending on the degree of interest and level of participants, the run chart could replace the control chart. It is evident that the greater the extent to which the process and the programme are developed, the more advanced the tools and techniques become. For example, a step change from Four or Five Sigma to Six Sigma, or moving to DFSS, would certainly require the use of advanced tools and techniques (Basu, 2009). These advanced tools and techniques should be considered for 'Black Belts' and expert members of the project team. Please also see Appendix 2: Introduction to basic statistics.

10.3 QUALITY PROGRAMMES

Knowledge of the tools and techniques is only one part of implementing a quality programme. Without a focused agenda the training activities will not deliver results. This will only create a situation similar to a group of trained craftsmen waiting for work; or it could be likened to a team of golfers hanging around at the 'nineteenth hole' without a tournament to play in. Whatever the name, objectives and scope of the programme may be, it is essential to have a structured plan in order to make the best use of your toolbox. We have adapted a FIT SIGMA programme from Basu and Wright (2003) to illustrate how to 'make it happen' and implement a quality course aided by the tools and techniques.

The implementation of FIT SIGMA is a major change programme designed to transform an organisation. Such a make-over can only come about if the cultural change to the prevailing mindset is combined with the transfer of facts. FIT SIGMA is not an ad hoc localised improvement project, it is rather a holistic programme spanning the whole organisation. Therefore, the essential characteristics of a FIT SIGMA implementation programme are:

1. Top led with totally committed management, and bottom driven.
2. Project management discipline of scope, time and budget.
3. Rigorous specialist training.
4. Company-wide open communication, spanning all functions.
5. Measurement of savings and success.

In the implementation of Lean thinking the pitfalls of a piecemeal approach should be considered carefully. When organisations adopt this sort of gradual strategy to implement the principles of Lean, their partial efforts often fail. If an organisation is fighting for survival then a focused but segmented approach for quick results is justified, but the recommended approach for a sustainable legacy must be across the total value chain.

In the preceding chapters we have already touched upon many factors related to implementation programmes. For example, Chapters 3 and 4 show the appropriate DMAIC methods and the tools and techniques of implementing FIT SIGMA. Chapters 5–9 include tips for implementing FIT SIGMA depending on the type and size of an organisation and the complexity of a problem. There are also many publications and articles relating to strategic change management and project management, for example Carnall (1999) and Turner *et al.* (1996), that would be helpful to read before embarking on a FIT SIGMA programme. Some other literature, however, implies that if an organisation follows a recommended systematic and structured methodology then change management is straightforward. Our experience shows that this sort of rigid, highly ordered approach is far from a guarantee of success. In what follows we recommend a number of proven pathways with some degree of flexibility, taking into account the requirements and existing culture of the organisation.

The proven pathways we outline are to implement FIT SIGMA for organisations that are in different stages of sigma awareness and development. We have categorised three stages of development:

1. New starters of FIT SIGMA.
2. Started Six Sigma, but stalled.
3. Completed Six Sigma, but where to now?

10.4 IMPLEMENTATION FOR NEW STARTERS

At this stage the management understand the need for change and the necessity for an improvement programme. The main concern will be the change required to the culture of the organisation and the absence of a proven structure for transformation of a culture. Management knows what they want but how do they convince the staff that they (the staff) need to or want to change? You can take a horse to water – but how do you make it drink?

Here we provide a total proven pathway for implementing a FIT SIGMA programme, from the start of the initiative to the embedding of the change, right through to a sustainable organisation-wide culture.

Note that the entry point, and the emphasis on each step of the programme, could vary depending on the 'state of health' of the organisation.

The framework of a total FIT SIGMA programme is shown in Figure 10.1 and described below.

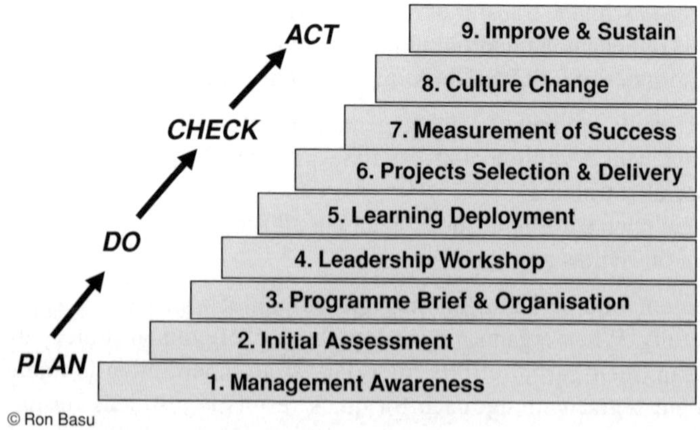

Figure 10.1 Framework of a total FIT SIGMA programme

10.4.1 Step 1: Management Awareness

A middle manager has been tasked by the CEO with leading a Six Sigma programme in a large organisation which has no previous experience of Six Sigma. The CEO has just read an article concerning Jack Welch's successes with Six Sigma at General Electric, and he is full of enthusiasm and has high expectations. The middle manager, however, is less enthused; in fact, he has grim forebodings of failure. He realises that the CEO is a powerful member of the board, but after all he is only one member. In another organisation the quality manager for a medium-sized company has attended a Six Sigma conference and has mixed feelings of optimism and some doubts. In fact, the author believes that both these managers are right to be concerned.

Our experience is that it is essential to convince the CEO and at least a third of the board regarding the scope and benefits of FIT SIGMA, prior to launching the programme. The success rate of a 'back door' approach without the endorsement of the key players cannot be guaranteed. If a programme is not company-wide and wholly supported by senior management it is simply not 'FIT SIGMA'. It may be a departmental improvement project – but it is not FIT SIGMA. In cricketing terms, a CEO can open the batting, but a successful opening stand needs a partner at the other end.

Research by the Corporate Leadership Council (2005) revealed that through using leadership and change training programmes, companies can substantially increase the potential for change initiatives to be successful. The research also indicated that companies that initiate too much change too quickly actually negatively affect the motivation of employees and their performance.

Our experience suggests that management awareness has been a key factor in the successful application of Six Sigma in large organisations. Various methods have been followed, including:

1. Consultant presentations to an off-site board meeting (e.g. General Electric).
2. The participation of senior managers in another organisation's leadership workshop (e.g. GSK and Ratheon).
3. Study visits by senior managers to an 'experienced' organisation (e.g. Noranda's visit to General Electric, DuPont and Alcoa).

Popular Six Sigma literature has advocated that Six Sigma cannot be implemented successfully without top management commitment (Pyzdek, 2003; Ladhar, 2007; Breyfogle, 2008). Moreover, Basu (2009) argues further that if the chief executive does not have a passion for quality and continuous improvement, and if that passion cannot be transmitted down through the organisation, then paradoxically the ongoing driving force will be from the bottom up. The experience of Novozymes A/S (see Chapter 11) demonstrated that top management did not commit to the initiative because they believed that dictating the means to achieve a goal was not compatible with Novozymes' culture. Ironically, the implementation of Lean Six Sigma was in fact supported by the culture of Novozymes with a long tradition for working with continuous improvements. It can be argued that the apparent lack of commitment but tacit support empowered the middle management and acted as an incentive to demonstrate tangible results.

Small and medium-sized firms can learn from the experience of larger organisations, and indeed there can be mutual benefits for the larger organisation through an exchange of fact-finding missions. A service industry organisation could well benefit by exchanging these sorts of visits with successful Six Sigma companies in the finance sector, such as American Express, Lloyds TSB and Egg plc.

During the development of the management awareness phase it is useful to produce a board report or 'white' paper summarising the findings and benefits. This account has to be well written and concise, but it should not be rushed. We recommend that you allow between four and 12 weeks for fact finding, including visits, and the writing of the 'white' paper.

10.4.2 Step 2: Initial Assessment

Once the agreement in principle from the board is achieved, we recommend an initial 'health check' should be carried out at the organisation. There are many good reasons for conducting an initial assessment before formalising a FIT SIGMA programme. These include:

1. Having a destination in mind, and knowing which road to take, are not helpful until you find out where you are to start.
2. You should get to know the organisation's needs through analysis and measurement of the initial size and shape of the business and its problems/concerns or threats. Once these are ascertained, then the techniques of FIT SIGMA can be tailored to meet these needs.
3. The initial assessment acts as a spring board by bringing together a cross-functional team and reinforces 'buy-in' at the middle management level.
4. It is likely that most organisations will have pockets of excellence along with many areas where improvement is obviously needed. The initial assessment process highlights these at an early stage.
5. The health check must take into account the overall vision/mission and strategy of the organisation, so as to link FIT SIGMA to the key strategy of the board. Thus, the health check will serve to reinforce or redefine the key strategy of the organisation.

There are two essential requirements leading to the success of the assessment (health check) process:

1. The criteria of assessment (check list) must be holistic, covering all aspects of the business, and specifically address the key objectives of the organisation.
2. The assessing team must be competent and 'trained' in the assessment process. (Whether they are internal or external is not a critical issue.)

It is sensible that the assessment team be trained and conversant with basic fact-finding methods, such as those used by industrial engineers. Some knowledge of the European Foundation for Quality Management (EFQM, 2003) would be most useful.

Once the health check assessment is complete, a short report covering strengths and areas for improvement is required. We stress that this report should be short (not the 75-page detailed account required for the EFQM). In writing the document the company might require the assistance of a Six Sigma consultant. The typical time needed for the health check is two to six weeks.

10.4.3 Step 3: Programme Brief and Organisation

This is the organisation phase of the programme requiring a clear project brief, the appointment of a project team and the development of a project plan. All elements are essential, since 'major, panic driven changes can destroy a company; poorly planned change is worse than no change' (Basu and Wright, 1998).

The brief must clearly state the purpose, scope objectives, benefits, costs and risks associated with the programme. A FIT SIGMA syllabus is a combination of TQM, Lean management, Six Sigma and culture change supervision. It is a huge undertaking and requires the disciplined approach of project management. 'Programme management is a portfolio of projects that change organisations to achieve benefits that are of strategic importance' (MSP, 2007).

One risk at this stage is that management might query the budget for the programme, and there might be some reluctance to proceed. If this is the case, then it is obvious that management has not fully understood the need for change. This is why we have stressed the importance of the first step, 'management awareness'. However, reinforcement could be needed during step 3, underpinned with informed assumptions and data including a cost/benefit/risk analysis. Unless management is fully committed, there is little point in proceeding.

There is no rigid model for the configuration of the FIT SIGMA team. Basic elements of a project structure for a major change programme can be found in Basu and Wright (1998). Our suggested FIT SIGMA model is shown in Figure 10.2.

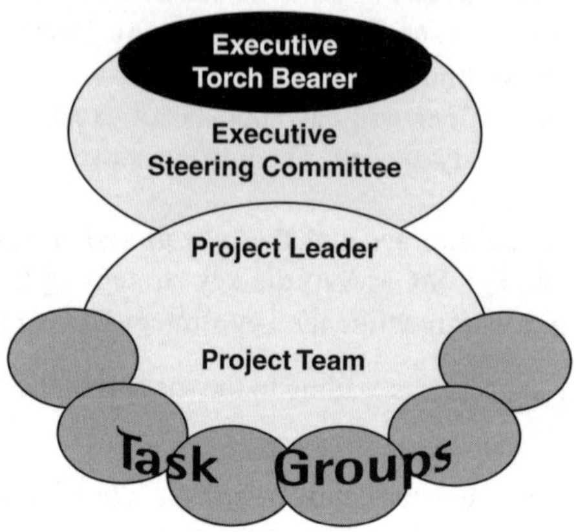

Figure 10.2 FIT SIGMA programme organisation

10.4.3.1 Executive Torch Bearer

Figure 10.2 shows an Executive Torch Bearer, who ideally will be the chief executive (CEO) and will be the official sponsor for FIT SIGMA. There is a correlation between the higher up the organisation the Torch Bearer is and the greater the success of the programme. The role of the Torch Bearer is to be the top management focal point for the entire programme and to chair the meetings of the Executive Steering Committee. Being a Torch Bearer may not be a time-consuming function, but it is certainly a very important role to give the programme a high focus, to expedite resources and to eliminate bottle necks.

10.4.3.2 Executive Steering Committee

To ensure a high level of commitment and ownership to the project, the steering committee should be drawn from members of the board plus senior management. Their role is to provide support and resources, to define the scope of the programme consistent with corporate goals, to set priorities and consider and approve the programme team recommendations. In Six Sigma terminology they are the champions of processes and functional disciplines.

10.4.3.3 Programme Leader

The Programme Leader should be a person of high stature in the company – a senior manager with broad knowledge of all aspects of the business and possessing good communication skills. He or she is the focal point of the project and also the main communication link between the Executive Steering Committee and the programme team. Often the Programme Leader will report direct to the Torch Bearer.

The Programme Leader's role can be likened to that of a consultant. The function of the leader is to a great extent similar to Hammer and Champy's 'czar' in *Re-engineering the Corporation* (1993). In other words, the Programme Leader's task is to:

- Provide necessary awareness and training for the project team, especially regarding multi-functional issues.
- Facilitate the work of various project groups and help them develop and design changes.
- Interface across functional departments.

In addition to the careful selection of the Programme Leader, two other factors are important in forming the team. First, the membership size should be kept within manageable limits. Second, the members should bring with them not only analytical skills but also an in-depth knowledge of the total business covering marketing, finance, logistics, technical and human resources. The minimum number of team members should be three, and the maximum should be seven. Any more than this can lead to difficulties in arranging meetings, communicating and keeping to deadlines. The dynamics within a group of more than seven people allows a pecking order to develop and for sub-groups to emerge. The team should function as an action group, rather than as a committee that deliberates and makes decisions. Their role is to:

- Provide objective input into the areas of their expertise during the health check stage.
- Lead activities when changes are made.

For the Programme Leader, the stages of the project include:

- Education of all the people in the company.
- Gathering the data.
- Analysis of the data.
- Recommending changes.
- Regular reporting to the Executive Steering Committee and to the Torch Bearer.

Obviously the Programme Leaders cannot do all the work themselves. A Programme Leader has to be the type of person who knows how to make things happen and who can motivate and galvanise other people to help achieve this aim.

10.4.3.4 Programme Team

The members of the Programme Team represent all functions across the organisation and they are the key agents for making changes. Members are carefully selected from both line management and a functional background. They will undergo extensive training to achieve Black Belt standards. Our experience suggests that a good mix of practical managers and enquiring 'high flyers' will make a successful project team. They are very often the process owners of the programme. Most of the members of the Programme Team are part-time. As a rule of thumb, no less than 1% of the total workforce should form the Programme Team. In smaller organisations the percentage will of necessity be higher, so that each function or key process is represented.

10.4.3.5 Task Groups

Task Groups are spin-off sets formed on an ad hoc basis to prevent the Programme Team getting bogged down in detail. A Task Group is typically created to address a specific issue. The topic could be relatively major, such as the Balanced Score Card, or comparatively minor, such as the investigation of losses in a particular process. By nature, the Task Group members are employed directly onto the programme on a temporary basis. However, by supplying basic information for the programme, they gain experience and Green Belt training. Their individual improved understanding and 'ownership' of the solution provide a good foundation for sustaining future changes and ongoing improvements.

10.4.3.6 Time frame

A preliminary time plan with dates for milestones is usually included in the programme brief.

We have now covered the first three 'Plan' steps shown in Figure 10.1.

10.4.3.7 The 'Do' Steps

In Figure 10.1 we can see that after the 'Plan' phase there comes the 'Do' phase (also see Figure 4.21).

Once the programme and project plan have been agreed by the Executive Steering Committee, they should receive a formal launch. It is critical that all stakeholders, including managers, employees, unions, key suppliers and important customers, are clearly identified. A high-profile programme launch targeted at stakeholders such as these is desirable.

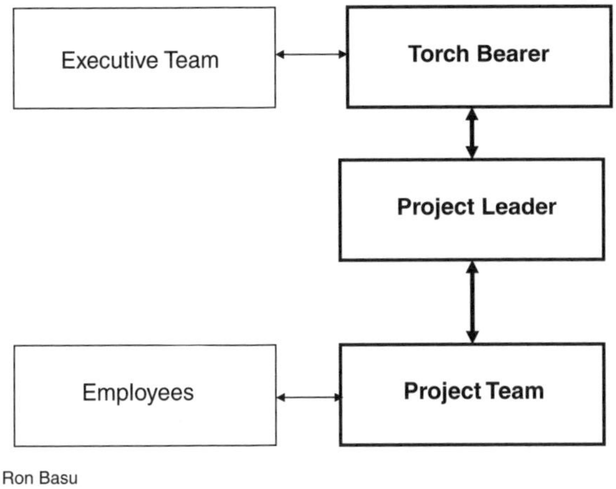

Figure 10.3 FIT SIGMA structure for SMEs

10.4.3.8 Organisation for Small and Medium Enterprises

The organisation structure of the programme will vary according to the nature and size of the organisation. For SMEs a typical structure is shown in Figure 10.3. In small enterprises the Programme Leader might be part-time. In all other cases the Programme Leader will be full-time.

10.4.4 Step 4: Leadership workshop

All board members and senior managers of the company need to learn about the FIT SIGMA programme before they can be expected to give their full support and input into the scheme. Leadership training is a critical success factor. Leadership workshops can begin simultaneously with step 1, but should be completed before step 5, see Figure 10.1. Workshops will last between two and five days and will cover the following issues:

1. What are Six Sigma and FIT SIGMA?
2. Why do we need FIT SIGMA?
3. What will it cost and what resources will be required?
4. What will it save, and what other benefits will accrue?
5. Will it interrupt the normal business?
6. What is the role of the Programme Leader and the Executive Steering Committee?

As shown in Chapter 11, leadership workshops or forums can play an important role in gaining acceptance for the change programme by members of both senior and middle management.

10.4.5 Step 5: Training Deployment

The training programme, especially for team members, is rigorous. One might question whether it is really necessary to train in order to achieve Black Belt certification. Indeed, formal certification might not be essential. However, there is no doubt that without the in-depth training of key members of the programme, little value will be added in the short term, and certainly not over a longer period. The training/learning deployment creates a team of experts. It is presupposed that programme members will be experts in their own departments and processes as they are currently being run. It is expected that they will have the capability to appreciate how the business as a whole will be organised in the future. FIT SIGMA will equip them with the tools for the business, as a whole, to achieve world-class performance.

Apart from the rigorous education in techniques and tools, it is emphasised that the training will change how the members look at things. Training is an enabler not only to understand the strategy and purpose of change but – as evidenced by the experience of American Express – it will help members to identify:

- Project replication opportunities.
- Leveraging the results of the programme.
- Identification and elimination of areas of rework.
- Drivers for customer satisfaction.
- Leverage of FIT SIGMA principles into new products and services.

Smaller organisations are very often concerned about the cost of training, especially the money paid out to consultants and for courses. In a FIT SIGMA programme, teaching costs can be minimised by the careful selection of specialist consultants and through the development of own in-house training programmes. Porvair Ltd is an example of a smaller enterprise which achieved good results with a limited training budget.

Case example: Porvair Ltd

Porvair is a manufacturing company based in Wrexham, England. It currently employs 80 people and the annual sales are £6 million. Its product is plastic, bronze or stainless steel, which is applied to porous media and filtration equipment. There are 2000 part numbers in the product range. The customers are involved in high-temperature catalyst recovery, medical applications, nylon spinning and water filtration.

Prior to introducing Six Sigma, the company had a respectable reputation – as demonstrated by the fact that the Welsh Development Agency designated the company as a benchmark site. However, the ad hoc improvement programme relied heavily on one person, the technical director. The other area of concern was that Porvair was experiencing poor delivery performance – less than 50% of deliveries were on time. Following a Six Sigma programme that commenced in May 2000, by March 2002 the company had achieved remarkable results, as tabled below.

Table 10.1 Porvair performance increase

	May 2000	March 2002
Delivery on time	less than 50%	90–95%
Head count	135	80
Customer complaints (per month)	12	7
Waste	14%	10%

Plus benefits from three Six Sigma projects of £206,000.

The company deployed a specialist Six Sigma consultant from Belfast. The costs of training courses were:

One Champion	£14,000
One Black Belt	£17,000
Four Green Belts	£14,000
Other costs	£12,000
Total	**£57,000**

The Black Belt (once trained) carried out further instruction in-house for additional Green Belts and awareness coaching for all employees. The training cost at Porvair equates to £1000 per employee per annum.

10.4.6 Step 6: Project Selection and Delivery

The project selection process usually begins during the training deployment step.

Project selection, and subsequent delivery, is the visible aspect of the programme. A popular practice is to begin by having easy, and well-publicised, successes (known as 'harvesting hanging fruit'). We recommend that quick 'wins' should be aimed for (or 'just do it' projects).

In a similar fashion, Ericsson AB applied a simplified 'Business Impact' model for larger schemes. They categorise ventures under three headings:

- Cost takeout.
- Productivity.
- Cost avoidance.

A variable weighting is allocated to each category, as shown in Table 10.2.

Business impact = Cost takeout + 0.5(Productivity) + 0.2(Cost avoidance) – Implementation cost

Table 10.2 Categories of savings

Level	Cost takeout	Productivity	Cost avoidance
Definition	'Hard' savings Recurring expense prior to Six Sigma Direct costs	'Soft' savings Increase in process capacity so you can 'do more with less', 'do the same with less', 'do more with the same'	Avoidance of anticipated cost or investment which is not in today's budget
Example	Less people to perform activity Less $ required for same item	Less time required for an activity Improved machine efficiency	Avoiding purchase of additional equipment Avoiding hiring contractors
Impact	Whole unit	Partial unit	Not in today's cost
Weighting	100%	50%	20%

Table 10.3 Derived importance of projects

Projects	Cost takeout High = 10 Low = 1	Productivity High = 5 Low = 1	Cost avoidance High = 2 Low = 1	Employee satisfaction High = 3 Low = 1	Current performance High = 1 Low = 10	Feasibility High = 10 Low = 1	Delivered importance Max score = 40
Design packaging	9	2	0	2	7	8	20
Improve OEE	5	4	2	2	10	10	33
Passes control	1	2	2	1	5	9	20
New product development	7	1	2	3	2	2	17

When sufficient data is not available to provide an accurate estimate of business impact, an approach of 'derived importance' based upon scores for various categories is a practical alternative, as shown in Table 10.3.

For smaller, 'just do it' projects, it is good practice to establish an 'ideas factory' to encourage Task Groups and all employees to contribute to savings and improvement. Very often small projects from the 'ideas factory' require negligible funding.

10.4.6.1 Project Review and Feedback

One important point of the project selection and delivery step is to monitor the progress of each task and control the effects of the changes so that expected benefits are achieved. The Programme Leader should maintain a progress register supported by a Gantt chart, defining the change, expected benefits, resources, time scale and expenditure (to budget), and show the people who are responsible for actions.

This phase of review and feedback involves a continuous need to sustain what has been achieved and to identify further opportunities for improvement. It is good practice to set fixed dates for review meetings, as follows:

Milestone Review (at least quarterly)
= Executive Steering Committee,
Torch Bearer and
Programme Leader

Programme Review (monthly)
= Programme Leader and Team
with a short report to Torch Bearer

The problems/hold-ups experienced during projects are identified during the Programme Review, with the aim of the Project Team taking action to resolve sticking points. If necessary, requests are made of the Executive Steering Committee for additional resources.

10.4.7 Step 7: Measurement of Success

The fundamental characteristic of a Six Sigma or FIT SIGMA programme that differentiates it from a traditional quality curriculum is that it is results-orientated. Effective measurement is the key to understanding the operation of the process, and this forms the basis of all analysis and improvement work. In a construction project the milestones are both tangible and physically obvious, but in a change programme such as FIT SIGMA the modifications are not always apparent. It is essential to measure, display and celebrate the achievement of milestones in a FIT SIGMA programme, and in order to improve and sustain its results we strongly emphasise the importance of performance management. The process and culture of measurement must start during the implementation of changes.

Our experience is that the components of the measurement of success should include:

- project tracking;
- FIT SIGMA metrics;
- Balanced Score Card;
- self-assessment review (Baldridge or EFQM).

10.4.7.1 Project Tracking

There are software tools available, such as Minitab (www.minitab.com) and KISS, for carrying out detailed tracking of larger Six Sigma projects. However, in most programmes the progress of savings generated by each project can be monitored on an Excel spreadsheet. We recommend that summaries of results are reported and displayed each month. Examples of forms of displays are shown in Figures 10.4 and 10.5.

10.4.7.2 FIT SIGMA Metrics

FIT SIGMA metrics are required to analyse the reduction in process variance and the reduction in the rate of defects resulting from the appropriate tools and methodology.

A word of caution: Black Belts can get caught up with the elegance of statistical methods and this preoccupation can lead to the development of a statistical cult. Extensive use of variance analysis is not recommended.

The following FIT SIGMA metrics are useful, easy to understand and easy to apply:

Cost of Poor Quality = COPQ ratio
Defects per Million Opportunities = DPMO
First Pass Yield = FPY

COPQ
[Internal failure ($) + External failure ($) + Appraisal and prevention ($) + Lost opportunity ($)] / Monthly sales ($)

DPMO
[Total number of defects $\times$ 1,000,000] / Total units and opportunities per unit

FPY
Number of units completed without defects and rework / Number of units started

From measuring and monitoring FIT SIGMA metrics each month, opportunities for further improvement will be identified.

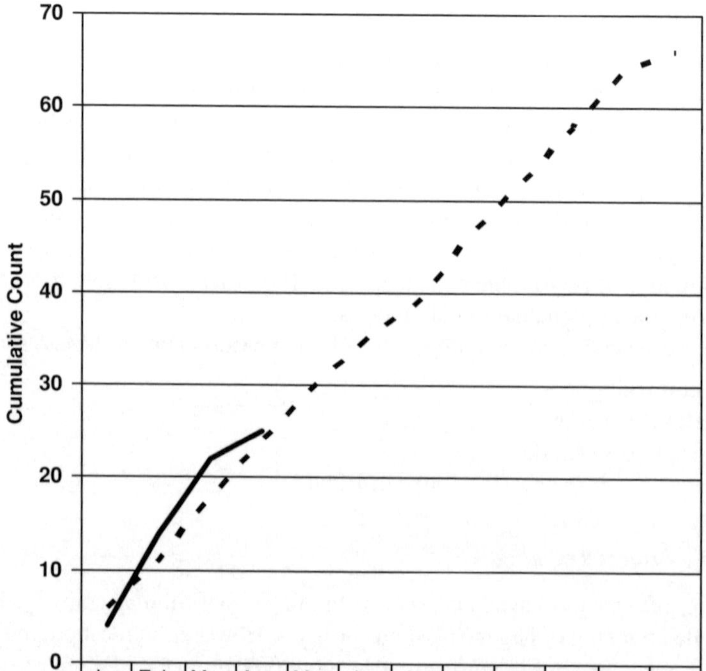

	Jan	Feb	Mar	Apr	May	Jun	Jul	Aug	Sep	Oct	Nov	Dec
Cum Act	4	14	22	25								
Cum Plan	6	11	18	24	30	35	39	46	52	58	64	66

Cum Act - - - Cum Plan

Figure 10.4 Projects planned and completed 2002

10.4.7.3 Balanced Score Card

As we have already emphasised, a carefully designed Balanced Score Card (see Chapter 2), is essential for improving and sustaining business performance. It is generally viewed that the Balanced Score Card is applicable for a stable process and thus should be appropriate after the completion of the FIT SIGMA programme. This may be so, but unless the measures of the Balanced Score Card are properly defined and designed for the purpose at an early stage, its effectiveness will be limited. Therefore, we strongly recommend that during the FIT SIGMA programme the basics of the Balanced Score Card to manage the company-wide performance system should be established.

10.4.7.4 Self-assessment and Review

There are two options to monitor the progress of the business resulting from the FIT SIGMA programme. Either of the following can be used:

1. A simple checklist to assess the overall progress of the programme.
2. A proven self-assessment process such as the European Foundation of Quality Management (EFQM) or the American Malcolm Baldridge system.

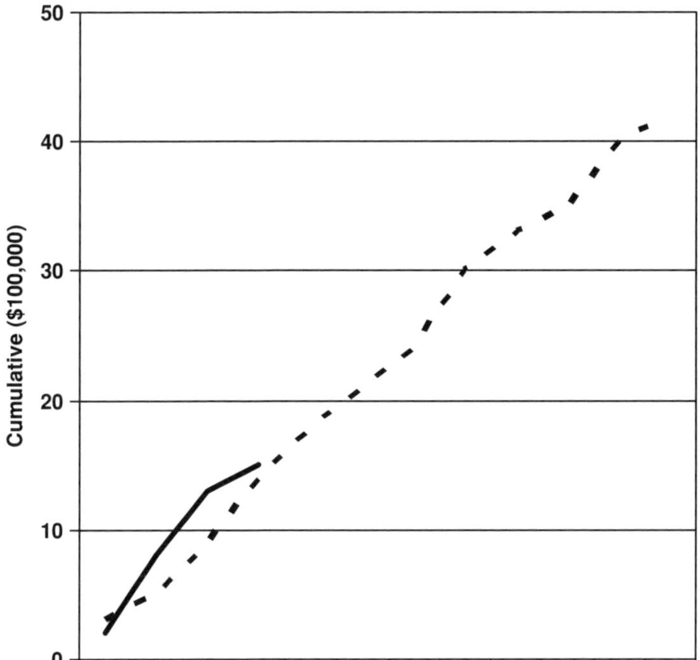

	Jan	Feb	Mar	Apr	May	Jun	Jul	Aug	Sep	Oct	Nov	Dec
Cum Act	2	8	13	15								
Cum Plan	3	5	9	14	18	21	24	30	33	35	40	42

© Ron Basu

Figure 10.5 Value of planned and completed projects

We recommend the second option. In the initial health check appraisal stage we advocate using EFQM or Baldridge, thus the methodology will already have been applied. Additionally, it gives further experience in the self-assessment process, which will enable future sustainability. Finally, it will provide the foundation should the organisation wish at a later stage to apply for an EFQM or Baldridge award.

10.4.8 Step 8: Culture change

A culture change must NOT begin by replacing middle management by imported 'Black Belts'. Winning over as opposed to losing middle management is essential to the success of FIT SIGMA, or for that matter any quality initiative.

What is required is that the all-important middle management, and everyone else in the organisation, understands what FIT SIGMA is, and possesses the culture of quality. An understanding of Deming's 14 points would be a sound basis from which to start (see Chapter 2).

The FIT SIGMA culture is shown in Table 10.4.

Table 10.4 FIT SIGMA culture

Total vision and commitment of top management throughout the programme.
Emphasis on measured results and the rigour of project management.
Focus on training with short-term projects and results, and long-term people development.
Use of simple and practical tools.
Total approach across the whole organisation (holistic).
Leverage of results by sharing best practice with business partners (suppliers and customers).
Sustaining improvement by knowledge management, regular self-assessment and senior management reviews.

Air Academy (www.airacad.com, April 2002) claim that it is important first to understand the culture type of an organisation to ensure the culture change necessary for the success of a Six Sigma programme. The culture styles they identify are:

1. Clan culture. The organisation is considered to be people-orientated. It is a nice place to work where people share similar interests, much like a country club, i.e. clan like – they all have similar beliefs and values.
2. Hierarchical culture. The organisation has a formalised top-down structure, and people are governed by rules and procedures.
3. Enterprise culture. The organisation is goal-orientated. Results are measured and members are competitive. People are primarily concerned with getting the job done.
4. Adhocracy culture. The organisation is dynamic, creative and entrepreneurial. People are proactive and take risks, are innovative and look for alternatives.

FIT SIGMA requires a balanced culture comprising key characteristics of the above four categories. If an organisation is predominantly one type then some cultural change will be required. Training deployment, see step 5 of Figure 10.1, includes preparation for culture change. But education alone will not transform the mindset required for FIT SIGMA.

10.4.8.1 Creating a Receptive Culture

A frequently asked question is how to change the prevailing culture. It all begins with vision . . .

That vision of quality must start with the chief executive. If the chief executive has a passion for quality and continuous improvement, and if this zeal can be transmitted down through the organisation, then paradoxically, the ongoing driving force will be from the bottom up rather than enforced from above, and with everyone sharing the same vision. For similar viewpoints regarding TQM see Crosby (1984), Ishikawa (1985), Schonberger (1986), Albrecht (1988) and Gabor (2000).

The word 'vision' might suggest almost a mystical occurrence (such as the visualisation experience of Joan of Arc), or an ideal (expressed by Martin Luther King in his famous 'I have a dream' speech). The same connotation is found when looking at the concept of 'vision' in an organisational context. A leader with a vision is a leader with a passion for an ideal, someone with purpose, moral values, strong beliefs and principles. 'But, unless the vision can happen, it will be nothing more than a dream (Wright, 1999, p. 20). To make a vision happen within an organisation, there has to be a cultural fit. Corporate culture is the amalgam of the existing beliefs, norms and values of the individuals who make up the organisation ('the way we do things around here'), according to Peters and Austin (1986). The leader may be the one who articulates the vision and makes it legitimate, but unless it mirrors the goals and aspirations of the members of the organisation at all levels, the ideal won't become a reality (Albrecht, 1988).

Culture and values are deep-seated and may not always be immediately apparent to members. As well as the seemingly normal aversion to change that many individuals possess, often there is a vested interest for members of an organisation to resist transformation. Middle management often is more likely to oppose the idea than are other members. Machiavelli (1513) wrote, 'It must be considered that there is nothing more difficult to carry out, nor more doubtful to success, nor more dangerous to handle, than to initiate a new order of things'. Human nature hasn't really changed much since the 16th century!

Organisations are made up of many individuals, each with their own set of values. The culture of each such society lies in how people react or do things when confronted with the need to make a decision. If the establishment has a strong culture, then each individual will instinctively know how things are done and what is expected. Conversely, if the corporate culture is weak, the individual may not react in the manner that management would hope (Peters and Austin, 1986; Carnall, 1999).

To engineer or change a culture there has to be leadership from the top. Guidance by example might seem to be a cliché but, unless the chief executive can clearly communicate and demonstrate a clear policy by illustration, how will the rest of the people know what is expected? Leadership does not have to be charismatic, but it has to be honest. Leadership does not rely on power and control. In fact, Basu and Wright (1998) find that real leaders communicate face to face, not by memos.

10.4.8.2 Mission Statement to Signal Change

A new mission statement would seem to be a logical way for a chief executive to signal a change in direction for an organisation.

Ideally, this mission statement should be a true avowal as to the organisation's reason for being. It should be realistic and state the obvious. Profit is not a dirty word (Friedman, 1990) and customer service is important (Zeithaml *et al.*, 1990; Kotler, 1999). Generally the key resource lacking in many organisations is quality people, a fact attested to by Barlett and Ghoshal (1994), Mintzberg (1996) and Knuckey *et al.* (1999). Therefore, it would seem obvious for any mission to say that they are in business to make a profit, that they will make that profit by providing the customers with what they want, and that they recognise that their most important resource in making this mission happen is their people. It is important that the new mission be in tune with what the people of the organisation believe (in other words, the existing culture). To achieve a purpose that fits the culture it would seem sensible to get the involvement and interest of all the staff in writing for the new mission. Thus, in this manner, a change in culture could begin with the determination and buy-in of staff into the new exercise.

10.4.8.3 Learning for Change

'If employees, organisation wide, are going to accept change, and themselves individually change, they will need to learn certain skills. Skills such as:

- understanding work processes,
- solving problems,
- making decisions, and
- working with others in a positive way.

All these types of skills can be taught. The main message that has to be learnt is the need for cultural change, and for people to trust each other. In particular management has to win the trust of lower level staff and have to learn how to change from autocratic management to coaching and mentoring. Lower level staff, in turn, have to learn to trust management' (Wright, 1999).

Please see also Hall (1999) and Axelrod (2001) who express similar views. Once this has been achieved, the culture will be such that the organisation will be in tune with the philosophy of FIT SIGMA, and the ninth step of FIT SIGMA, 'improve and sustain', will be second nature.

10.4.8.4 Communication

Finally, the key to sustaining a FIT SIGMA culture is the process of good communication. Methods of communication include:

- An intranet FIT SIGMA website, specifically developed, or clearly visible on the corporate website.
- Specially produced videos, or CDs.
- A FIT SIGMA monthly newsletter.
- Internal emails, voicemails, memos with updated key messages – NOT slogans such as 'work smarter not harder' and other tired clichés.
- Milestone celebrations.
- Staff get-togethers, such as special morning teas, a Friday afternoon social hour, or 'town hall'-type meetings.
- An 'ideas factory' or 'think tank' to encourage suggestions and involvement from employees.

10.4.9 Step 9: Improve and Sustain

'Improve and sustain' is the cornerstone of a FIT SIGMA programme. This is similar to Tuckman's (1965) fifth stage of team dynamics for project teams ('Forming, Storming, Norming, Performing and Mourning'). During the Mourning step the project team disbands and members move onto other ventures or activities. They typically regret the end of the project and the break up of the group, and the effectiveness or maintenance of the new method and results gradually diminish. In Chapter 6 we discussed in some detail that in order to achieve sustainability, four key processes must be in place:

1. Performance management.
2. Senior management review.
3. Self-assessment and certification.
4. Knowledge management.

The 'end game' scenario should be carefully developed long before the completion of the programme. There may not be a sharp cut-off point like a project handover and the success of the scenario lies in the making of a smooth transition without disruption to the ongoing operation of the business.

As part of the performance management, improvement targets should be gradually, and continuously, stretched and more advanced tools considered for introduction. For example, the DFSS (Design for Six Sigma) is resource-hungry (Basu, 2009), and can be considered at a later stage in a FIT SIGMA programme. With Six Sigma, the aim is to satisfy customers with robust 'zero-defect' manufactured products. In order to do so, DFSS is fully deployed covering all elements of manufacturing, design, marketing, finance, human resources, suppliers and key customers (including the supplier's suppliers and the customer's customers).

At an advanced stage of the programme, milestone review should be included in senior management operational review team meetings (such as the sales review meetings and operational planning meetings/committees). In other words, milestone review should occur not only within the FIT SIGMA Executive Steering Committee.

We recommend that a pure play EFQM (or other form of self-assessment) should be incorporated as a six-monthly feature of the FIT SIGMA programme. Even if the company gains an EFQM or Baldridge award, the process must still continue indefinitely.

Two specific features of knowledge management need to be emphasised. First, it is essential that the company seek leverage from FIT SIGMA results by rolling it out to other business units and main suppliers. Second, it is equally important to ensure that career development and reward schemes are firmly in place to retain the highly trained and motivated Black Belts. The success of the sustainability of FIT SIGMA occurs when the culture becomes simply an undisputed case of 'this is the way we do things'.

10.4.9.1 Time Scale

The time scale of FIT SIGMA implementation will last several months and is, of course, variable. The duration depends not only upon the nature or size of the organisation, but also on the business environment and the resources available. Four factors can favourably affect the time scale:

1. Full commitment of top management and the board.
2. Sound financial position.
3. Correct culture (workforce receptive to change).
4. A competitive niche in the marketplace.

It is good practice to prepare a Gantt chart containing the key stages of the programme and to monitor progress. Figure 10.6 shows a typical timetable for a FIT SIGMA agenda in a single-site medium-sized company. The diagram shows an order of magnitude only, and the sequence could well vary. The time line is not linear, stages overlap, and frequent retrospection should occur in order to learn from past events and work towards future progress.

10.4.9.2 FIT SIGMA for 'Stalled' Six Sigma

Some organisations have already attempted to implement a Six Sigma (or TQM) programme, but the process has stalled. Results are not being achieved and enthusiasm is waning; in some cases the programme has effectively been abandoned. The reasons for stalling are various, but often progression has ground to a halt due to an economic downturn (such as that experienced in the telecommunications industry in 2001); a change in top management; or a merger or take-over.

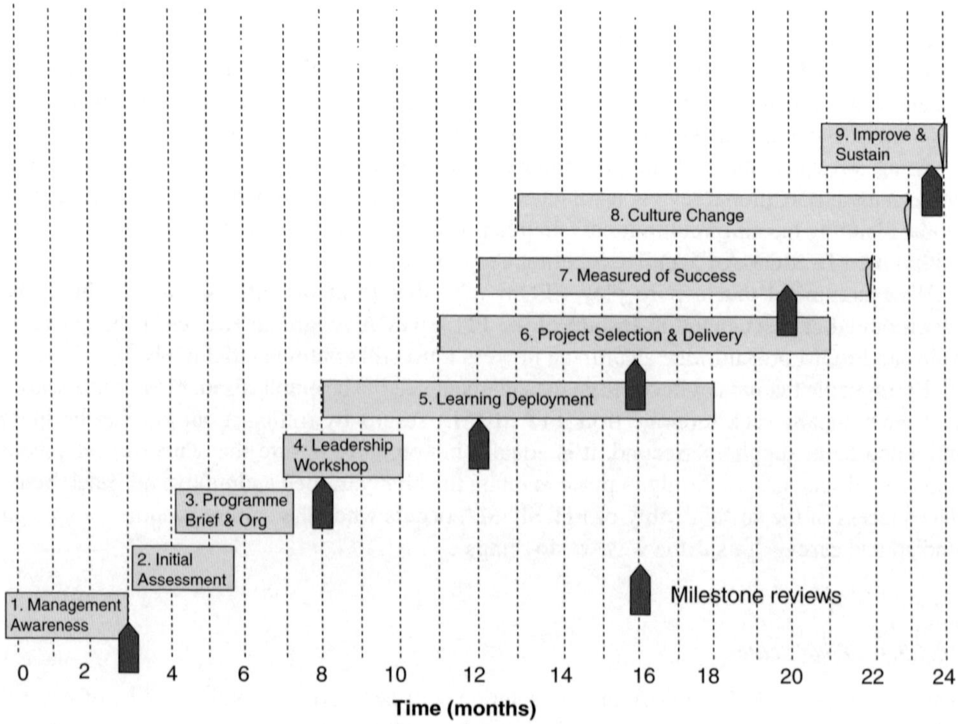

Figure 10.6 A typical time plan for a FIT SIGMA programme

10.4.9.3 FIT SIGMA: Not a 'quick fix'

During restructuring, or if the company is in survival mode, the implementation of FIT SIGMA is not appropriate. FIT SIGMA is not simply a 'quick fix', a plaster over gaping wounds, and in such cases it is not sufficient. Instead the underlying, more serious, issues must be addressed. After the short-term cost saving measures of a survival strategy, when the business has stabilised and a new management team is in place, *then* Six Sigma can be restarted, but this time it should be done correctly, and using the FIT SIGMA approach.

It is likely, when restarting, that many of the steps will not need to be repeated, including training/learning deployment. However, in a restart there is one big issue that makes life more difficult, and that is credibility. How do you convince all the people that it will work the second time around, when things didn't come together at the first attempt? This will put special pressure on step 8, culture change. The employees could well be tired of excessive statistics and complex Six Sigma tools. Thus, the selection of appropriate tools is a strong feature of FIT SIGMA.

The FIT SIGMA programme for a restart will naturally vary according to the condition of the organisation, but the programme can be adapted within the framework shown in Figure 10.1. The guidelines for each step are:

1. Management awareness. If not, then you cannot restart.
2. Initial assessment. This has to be the restart point. Where are we, where do we want to go?

3. Programme brief and organisation. The programme will need to be re-scoped and new teams formed.
4. Leadership workshop. This will be essential, even if management has not changed.
5. Training/learning deployment. Appropriate tools should be selected. If past team members, in particular Black Belts, are not happy with the terminology of the old programme, then new expressions should be used. The title 'Black Belt' in itself is not sacrosanct and might be changed. If the 'old' experts are still in the organisation, then training time might be reduced, i.e. only a one-week workshop might be sufficient.
6. Project selection and delivery. Same as the full FIT SIGMA programme, harvest the 'hanging fruit'.
7. Measurement of success review. Looking at the old measures, ascertaining what worked, what didn't, and following the full FIT SIGMA programme.
8. Culture change. This will be critical. Top management support will have to be extremely obvious. Reward and appraisal systems will have to be aligned to FIT SIGMA.
9. Improve and sustain. Same as for FIT SIGMA.

10.5 FIT SIGMA FOR SUCCESSFUL COMPANIES

William Stavropoulos, the CEO of Dow Chemical, is reported to have once said, 'The most difficult thing to do is to change a successful company'. It is true that employees of firms enjoying a high profit margin with some dominance in the market are likely to be complacent and to feel comfortable with the status quo. Perhaps it is even more difficult to remain at the top or to sustain success if the existing strategy and processes are not adaptable to change. Darwin famously observed, 'It is not the strongest species that survive, nor the most intelligent, but the ones most responsive to change'. It is possible for the management of some companies, after the completion of a highly successful Six Sigma programme, to find their attention diverted to another major initiative such as e-Business or Business-to-Business Alliances. Certainly new schemes must be pursued, but at the same time the long-term benefits that could be achieved from Six Sigma should not be lost. FIT SIGMA for sustainability – staying healthy – is the answer.

If a company has succeeded with Six Sigma, then the time is now right to move onto FIT SIGMA to achieve step 9, improve and sustain.

10.6 EXTERNAL CONSULTANTS

Many businesses, especially SMEs, are often concerned with the cost of consultants for a Six Sigma programme. Large consulting firms and academies for Six Sigma could well expect high front-end fees. However, with FIT SIGMA our approach is to be selective in the use of outside consultants. We advocate that you should use outside consultants to train your own experts, and to supplement your own expertise and resources when necessary. No consultant will know your own company as well as your own people will. For this reason we do not favour the use of external consultants in the role of Programme Leader.

In a FIT SIGMA programme the best use of consultants is in:

- Step 2: initial assessment. Here we would use a Six Sigma expert or an EFQM consultant to train and guide your team.

- Steps 4 and 5: leadership workshop and training/learning deployment. Outside consultants will be needed to facilitate the leadership workshops, and to tutor your own 'Black Belts'. Once trained, your own 'Black Belts' will in turn train 'Green Belts' and develop new 'Black Belts'.
- Step 8: culture change. An outside consultant is best suited to develop a change management plan for change of culture.

10.7 SUMMARY

This chapter provides practical guidelines for selecting appropriate tools and techniques and 'making it all happen' in a quality programme like FIT SIGMA. Many a Six Sigma exercise started with high expectations and looked good on paper. Many an organisation has been impressed by success stories of Six Sigma, but unsure of how to start. The implementation plan shown here will enable any organisation at any stage of a FIT SIGMA initiative to follow a proven path to success and to sustain benefits. Our implementation plan has nine steps, beginning with management awareness right through to the ongoing process of improve and sustain. There is no end!

In the spirit of FIT SIGMA, fit for purpose, this framework can be adjusted and customised to the specific needs of any organisation. Explicit comments have been included in this chapter for the implementation of FIT SIGMA in small and medium enterprises, where resources are constrained. Instead of all nine steps in Figure 10.1, SMEs should focus primarily on steps 1, 3, 5, 6 and 9.

At all stages of the programme, it is essential not only that the Executive Steering Committee and the Torch Bearer are kept informed (the Torch Bearer will keep the board up to date), but that there is open communication with all members of the organisation, so that everyone is aware of the aims, activities and successes of the programme.

11

More Case Examples

11.1 INTRODUCTION

The first wave of Six Sigma, following the impressive groundwork carried out by Motorola, included Allied Signal, Texas Instruments, Ratheon and Polaroid (to name a few). GE entered the arena in the mid-1990s and in turn was followed by numerous powerful corporations including SONY, DuPont, Dow, Bombadier, Ford and GSK. The success of GE Capital attracted many companies in the service sector, such as American Express, Lloyds TSB, Egg.com and Vodafone, to apply Six Sigma as a major business initiative. Some established companies like Unilever, ICI and BT have also applied Six Sigma tools and techniques but attempted to align them to existing strategic and continuous improvement programmes.

The ability to leverage the experience of successful Six Sigma players proved highly attractive, both as a competitive issue and also to improve profit margins. The following case examples provide insights for organisations that are achieving success in their business performance through the use of Six Sigma or related operational excellence programmes. The objective of including these case examples is to offer a set of good proven practices through which a student or practitioner can obtain a better insight into the application of tools and techniques and appreciate how the organisations have adopted their approach.

These case examples are taken from the real-life experiences of many organisations around the world. The examples reflect projects within respective operational excellence programmes. We have not made any effort to tidy them up or present them in a more glamorous light just to suit a certain subject area. It is hoped that these cases provide a useful resource to stimulate discussions and the training of tools and techniques in quality management.

11.2 CASE EXAMPLES FOR LARGE MANUFACTURING ORGANISATIONS

The case examples are presented in two categories: applications in large manufacturing and non-manufacturing organisations. The examples of large manufacturing organisations in this section show variations in approach depending on the complexity of problems involved.

Case Study 1: Product Innovation at Elida Faberge, UK

Elida Faberge, UK is a subsidiary of the Unilever Group and manufactures and supplies personal care products like shampoo, deodorants and toothpastes. The company examined the way in which innovation takes place in the business and how it is managed and controlled. The purpose of this review was to develop a clearly mapped transparent process for new products delivery ahead of competition, with the optimum use of resources. The re-launch of a major shampoo brand (Timotei) in March 1992 was the pilot project using this process.

The re-launch involved four variants of shampoo in four sizes and two types of conditioner in three sizes. It also incorporated the main technical changes of a new dispensing cap and alterations to its formulations.

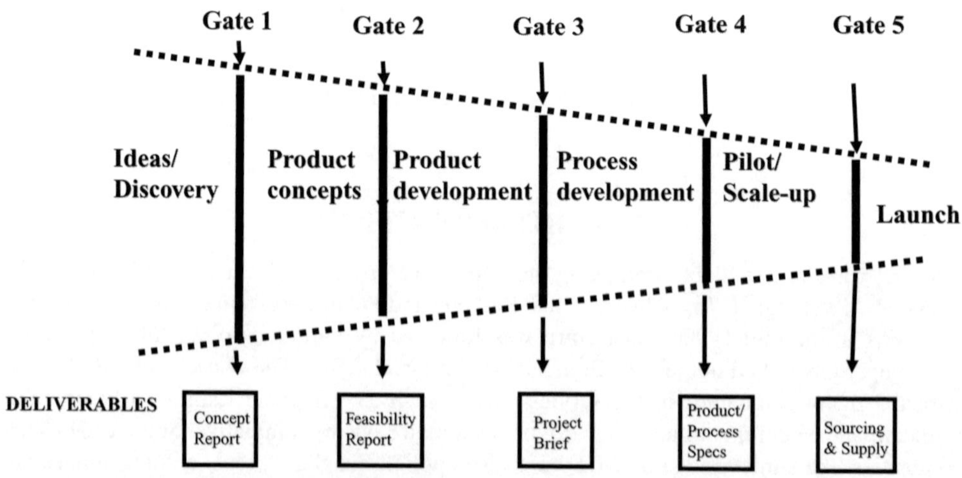

Figure 11.1 Product development phases and 'gates'

A multi-disciplinary task force was set up to develop the new process in readiness for a company-wide launch in April 1993. The innovation process used by Elida Faberge was based on the principles of 'phases and gates' (see Figure 11.1) and consisted of:

- A logical sequence of phases and exit gates.
- The use of empowered multi-disciplinary core teams.
- Using gates as key points where 'gatekeepers' review progress and decide whether a project may proceed or be terminated.
- Monitoring progress at each gate and for the overall project using key performance measures.

The project team had a membership representing the key processes, such as brand development, supply chain, quality management and customer development. The members received additional training in project management and the use of SPC tools. A project leader co-ordinated the team's efforts and reported the progress to the 'gatekeepers'.

Using the multi-functional team approach as described, the Timotei project was launched and managed to the implementation stage (Phase 4) by a team led by a brand manager. A second group was set up under the technical leadership of a development manager to see the project through to launch; in particular to set up supply lines, production trials and production start-ups.

Two other teams were also involved – a separate packaging supply chain team and a TQ team on the Timotei start-up production line. Technical audits were also established at three packaging supply sites and corrective actions were issued. These sites were revisited to ensure that all corrective actions had been satisfactorily achieved.

Results and Learning Points

The Timotei project was a success both in terms of sales and as a managed innovation process. In fact, no notable problems with materials supply were encountered. However, the considerable sales success of the product overstretched the bottle supplier, so that a second manufacturer was brought on stream. The quality and performance ratings of the production line achieved their targets.

The other learning points of the project included:

- The use of a multi-functional team allowed actions to be taken swiftly with shared ownership and resources at the team's disposal.
- The additional training of team members in project management and SPC tools ensured that the project was managed in a systematic, organised and disciplined manner.
- The effective use of phases (with multiple project leadership) and gates contributed to the development of the Unilever 'innovation funnel', which was introduced to all product groups.

Source: Implementing Six Sigma and Lean (Basu, 2009).

Case Study 2: Lean Manufacturing at an Aluminium Industry, USA

Author and researcher Dr James P. Womack, president of the Lean Enterprise Institute, told attendees at a meeting of the Aluminium Association that aluminium producers are among the many basic industries that could do a lot more with less difficulty and lower their costs if they did some lean thinking.

He described the labyrinthine and costly processes associated with delivering an aircraft, automobile or drinks can, and suggested manufacturers think about the consumer first and work backward to gain efficiencies and to cut out the non-value-adding activities.

'Ask, from the customer's standpoint, what is of value among your activities? What is wasted? How can we eliminate the waste? It's so simple and yet so very hard to do. Most people are in love with their assets, technologies and organisation', Womack said. Womack cited a military aircraft programme as an example. 'The typical subassembly goes through four plants, four states and 74 organizational handoffs between engineering, purchasing and fabricating operations. It goes 7,600 miles and takes darn near forever. It takes two to four years from beginning to flyaway condition.' Manufacturers need to identify the value stream for each of their products, and document all the steps it takes to go from raw material to the customer.

'Step 1, get the value right', said Womack. 'Your customer is not interested in your assets. He is interested in his value. Step 2, identify the value stream from start to finish, not just within the walls of your plant or company.'

Organise the remaining value-creating steps so they impact the product in a continuous flow, he advised.

Results and Learning Points

'If, and only if, you can create flow, then you can move to a world of pull', Womack said. 'You put the forecast in your shredder and get on with your life. And you make people what they want. The customer says, "I want a green one". And you say, "Here's a green one". That's a very different world from your world of endless forecasts, always wrong, and the desperate desire to keep running, which makes you produce even more of the wrong thing because that makes the numbers look good in the short term.'

Source: Metal Center News, May 1999, Volume 39, Issue 6, p. 123.

Case Study 3: Supply Chain Logistics at National Starch, USA

The National Starch and Chemical Company, based in Chicago, began measuring shipping performance on a corporate basis in the early 1990s. The initial results indicated that, at best, the Adhesives Division was achieving 80–85% on-time delivery performance. This is a business where it is essential to ship a multitude of products in a variety of containers with varying lead times from customers. The logistics department set an objective to reduce shipping delays from 15% to 5% and at the same time reduce overall inventories.

The approach the division took was to classify their products in specific containers by categories A, B and C. The categories represented volume of product and the customer base purchasing the product. The lead time for shipping the product and the inventory held were related to the A, B and C designations, with A being a shorter lead time with available inventory.

A number of local cross-functional teams were formed comprising sales, manufacturing, technical service, materials management and customer service. The task of the teams was to work on ABC product classification and delivery service standards. Essentially every product in each container was evaluated based on the volume and number of customers purchasing it. The Chicago plant reviewed over 200 products in approximately 15 different containers for more than 700 orders per month.

The next task was to develop a tracking system to report on delays against standard lead times and compare them with the date agreed with the customer.

After agreement was reached on the ABC products and their safety stocks, the order patterns of these products were reviewed and minimum, maximum and reorder levels were established. Employees were thoroughly trained and the system was also presented at each regional sales meetings. It was important to gain the support of sales staff since they needed to 'sell' it to the customers.

To ensure that the ABC and safety stocks reflected changing market demands, a cross-functional team reviewed the product A, B, C designations and inventory lists every month.

Results and Learning Points

Since the implementation of the ABC system, the improvement in shipping performance was dramatic, as shown by the figures for shipping delays as follows:

Year 0 overall:	14.7%
Year 1 overall:	3.8%
Year 2 overall:	1.2%

There were other benefits in the supply chain operations and performance, including:

- The sales team increased the amount of time used to work proactively with customers.
- There has been a reduction in finished goods and the raw materials inventory.
- Decreased 'last-minute' expedited freight charges.
- Improved communication between departments, especially between sales and logistics.

Source: Implementing Six Sigma and Lean (Basu, 2009).

Case Study 4: Performance Monitoring at Dupont Teijin Films, UK

Dupont Teijin Films is a global polyester films business with manufacturing sites in the USA, Europe and Asia. The company was created following the acquisition of Teijin Films of Japan by Dupont. DTF is a market leader but was experiencing tough competition from new entrants. As part of the corporate Six Sigma programme, the Wilton site of DTF in Middlesbrough, UK started the deployment plan from 1999.

The main objectives of the programme included:

- increased capacity;
- improved material efficiencies;
- cost reduction;
- increased revenue by higher sales volume.

The site project team followed a methodology of 'successful implementation' in three key categories: 'doing the right work' (Process); 'doing the work right' (Efficiency); and 'creating the right environment' (Education and Culture).

Within the category of 'doing the work right', the team introduced:

- input metrics;
- output metrics;
- tracking profile.

The input metrics included the number of Black Belts trained and people educated. The output metrics covered:

- $ saved;
- number of projects per annum;
- quality index;
- CTQ flowdown;
- COPQ;
- strategic linking.

The project team followed an internal self-assessment process every quarter based on a 'do right work checklist' comprising 24 questions in customer alignment, business alignment, process base lining and project selection.

Results and Learning Points

The Six Sigma programme has been running at the Wilton site since 1998. The typical sources and relative magnitude of savings are given in the following table:

Source	Year 1	Year 5	Relative value of savings
'Low-hanging fruits'	40%	5%	$$
COPQ	40%	20%	$$$$
KPIs	10%	10%	$$
CTQ flowdowns	10%	30%	$$$$
Strategic linking	0%	35%	$$$$$

The key learning point from DTF at Wilton is that training is a key deployment opportunity to influence the hearts and minds of your people. The structured continuous training programme has been vital to sustaining the Six Sigma programme for over five years.

Source: Implementing Six Sigma and Lean (Basu, 2009).

Case Study 5: Total Productive Maintenance at Nippon Lever, Japan

The Utsunomiya plant in Japan was commissioned in 1991 on a greenfield site by Nippon Lever to manufacture household detergents products and plastic bottles for liquid detergents. The factory was experiencing 'teething' problems primarily due to the poor reliability and lack of local support of the imported equipment. Many of the employees were new to factory work.

To improve this situation, the company used the help of the Japanese Institute of Plant Maintenance (JIPM), an organisation which was working on TPM (Total Productive Maintenance) with over 800 companies in Japan. TPM has been widely used in Japan, having been developed to support Lean/JIT and TQM. It was considered to be appropriate for the Utsunomiya plant since TPM focuses on machine performance and concentrates on operator training and teamwork.

A TPM programme was launched at the Utsunomiya plant in July 1992 with the objective of zero losses:

- zero stoppages;
- zero quality defects;
- zero waste in materials and manpower.

Strong organisational support was provided by the Nippon Lever management in terms of:

- A top management steering team to facilitate implementation by removing obstacles.
- A manager to work full time supporting the programme.
- One shift per week set aside for TPM work.
- Training for managers, leaders and operators involving JIPM video training material.

The programme launch was initiated at a 'kick-off' ceremony in the presence of the whole Nippon Lever board and managers from other companies and suppliers' sites.

The initial thrust of the programme was the implementation of 'autonomous maintenance' following the JIPM's seven steps:

1. Initial clean-up.
2. Elimination of contamination.
3. Standard-setting for operators.
4. Skill development for inspection.
5. Autonomous inspection.
6. Orderliness and tidiness.
7. All-out autonomous working.

To implement the seven steps, 'model machines' (those giving the biggest problems) were chosen. This approach helps to develop operators' knowledge of a machine and ensures that work on the model can be used as the standard for work on other machines. It also enhances motivation, in that if the worst machine moves to the highest efficiency, this sets the tone for the rest of the process.

The improvements to the machines were made using Kaizen methodology (small incremental improvements), and were carried out by groups of operators under their own guidance. Two means of support were given to operators – a Kaizen budget per line so that small repairs and capital expenses could be agreed without delay, and the external JIPM facilitator provided encouragement and experience to workgroups.

Results and Learning Points

By the end of 1993, substantial benefits were achieved within a year at the Utsunomiya plant, including:

- £2.8 million reduction in operating costs.
- Reduced need for expensive third-party bottles.
- Production efficiency increased from 54% to 64% for high-speed soap lines and from 63% to 80% for liquid filling lines.
- A team of trained, motivated and empowered operators capable of carrying out running maintenance.

The success of the programme at the Utsunomiya plant led to the introduction of TPM in two other factories at Nippon Liver (Shimizu and Sagamihara). Over the next few years the Corporate Groups of Unilever encouraged all sites outside Japan to implement TPM with remarkable successes achieved, particularly in factories in Indonesia, Brazil, Chile, the UK and Germany.

Source: Implementing Six Sigma and Lean (Basu, 2009).

Case Study 6: Six Sigma Training at Noranda, Canada

Noranda Inc. is a leading international mining and metals company for copper, zinc, magnesium, aluminium and the recycling of metal. With its headquarters in Toronto, the company employs 17,000 people around the world and its annual turnover in 2000 was $6.5 billion.

In August 1999, the board of Noranda decided to embark upon a global Six Sigma project with an initial savings target of $100 million in 2000. There were some specific challenges to overcome. The company business was in a traditional industry with long-serving history. Furthermore, Noranda is a 'decentralised' company with multiple cultures and languages. Senior executives studied the experiences of other companies (GE, Allied Signal, Dupont, Bombardier and Alcoa) and invited the Six Sigma Academy from Arizona to launch the training deployment programme.

The Six Sigma structure at Noranda focused on the training of the following levels:

- deployment and Project Champions;
- Master Black Belts;
- Black Belts;
- business analysts and validates;
- process owners;
- Green Belts.

The Six Sigma Academy was intensely involved for the first three months of the programme and then Noranda started its own education and training.

Results and Learning Points

The training accomplishments in 2000 were impressive:

- All 84 top executives followed a two-day workshop.
- 90 Black Belts were certified.
- 31 Champions were trained.
- There were 17 days of Master Black Belt training.
- 3000 days of Green Belt training.
- More than 3500 days of training in 2000 – and this has continued.

The key learning points from the experience of Noranda include:

- The rigorous deployment of structured learning schedules is key to the success of a company-wide Six Sigma programme.
- After the initial learning by external consultants, it is more cost-effective and useful in the longer term to build up your own in-house training teams.

Source: Implementing Six Sigma and Lean (Basu, 2009).

Case Study 7: Total Quality at Chesebrough-Pond's, Jefferson City

The Jefferson City Chesebrough-Pond's plant was built in 1966 and managed through the traditional functional structure. Each packing department had its own manager and supervisors responsible for each shift, resulting in five reporting layers. Maintenance and industrial engineering were separate departments and production planning was located with accounts in the office area. Raw materials and packaging suppliers were managed by the purchasing department in the head office at Connecticut, 1000 miles away.

With the concentration of the purchasing power of the retail trade in the USA (e.g. Wal-Mart, K-Mart, etc.), the plant had come under considerable pressure to deliver personal products more often, at shorter lead times and at a lower cost. Therefore, the factory decided it had to increase flexibility while at the same time reducing operating costs.

The management of the Jefferson City plant recognised that an organisational change was required to meet the demands of the business, and launched its total quality programme in the early 1990s.

The process started with the writing and adoption of statements on plant vision, plant mission and plant values. These statements were communicated throughout the plant. The other key activities in establishing total quality at the works were:

- understanding the supply chain;
- process management;
- problem-solving skills;
- setting goals and measuring progress;
- training.

The Jefferson City plant displayed a plant vision, mission and values by posting them in the permanent (yearly) framed charts in primary production buildings.

The plant was divided into three focused manufacturing units replacing the traditional functional management structure. All employees were referred to as 'associates' and participated in self-directed workgroups. The first-line supervisors were redeployed as 'coaches' of workgroups formed by associates. Each of these workgroups owned activities such as supplier contract, production scheduling and small projects.

All associates in the plant received at least six days' training per year in total quality tools and principles. Topics included quality characteristics, SPC, GMP, safety, scheduling and waste analysis. Even supplier employees visited the plant for four days' training in TQ.

Results and Learning Points

Within three years the plant achieved the satisfying of customer expectations through improved quality and matching production on the basis of demand and lower costs. For example:

- First pass quality improved from 93% to 99%.
- Work-in-progress inventory reduced by 86% and total inventory by 65%.

As a result, Chesebrough-Pond's at Jefferson City was recognised as one of its ten best plants in the USA by *Industry Week* magazine in 1992.

Source: 'Our search for the best', *Industry Week*, USA, October 1992.

Case Study 8: Sustaining FIT SIGMA by Leadership Forums at Pliva, Croatia

Pliva is a generic pharmaceutical company with manufacturing sites in Croatia, Poland and the Czech Republic. It has a broad range of product portfolio, including *Sumamed.* During the summer of 2004, Pliva embarked upon a Six Sigma programme called PEP (Pliva Excellence Process) and retained an American consulting company to train Black Belts and initiate projects. Although a good foundation for training 16 Black Belts was accomplished, underpinned by DMAIC-based projects, a change of mindset and culture was lacking. In 2005, Performance Excellence Limited was engaged to re-launch the programme. The primary focus of this second wave included the certification of Black Belts with completed projects, the training of additional Black Belts and Green Belts, and the management buy-in by using leadership forums. The following constitutes a detailed account and outcome of one such forum.

PEP came of age. Since the launch of this Six Sigma initiative in Global Product Supply (GPS) in the summer of 2004, PEP demonstrated tangible results as reflected by the projected savings of US$13.4 million, of which $4.6 million has already been realised. Perhaps more significantly, PEP opened the door towards a self-supporting sustainable operational excellence competence and holistic culture in most business units of GPS, supported by 28 trained Back Belts and 108 qualified Green Belts. The time was right to leverage this success and good practice across and beyond GPS. A most critical success factor for a major change programme such as PEP was the better understanding of the process by, and commitment of, managers. With this backdrop in mind, the first PEP leadership forum was conducted by Zeljko Brebric, PEP director, supported by Ron Basu, the retained consultant and mentor, and the PEP team.

The event took place in Zagreb and was attended by 24 managers from Pharma Chemicals, Local Product Supply Zagreb, Finance, Regulatory Affairs, Business Development, Medical Marketing Services and Human Resources. The primary objectives of the forum were to engage front-line managers in understanding Six Sigma/FIT SIGMA and PEP and to identify data-driven projects.

Zeljko Brebric presented an update of PEP and emphasised that PEP would be a continuous journey of achieving and sustaining operational excellence. The present strength of Black Belts and Green Belts would be fortified by additional stages of training and the next wave of in-house Black Belt education would soon commence. Ron Basu then explained the quality movement leading to Six Sigma and FIT SIGMA and what they meant to Pliva. This was followed by presentations by Ron Basu and interactive discussions on DMAIC methodology, project selection, Six Sigma tools and how to make it happen. It was evident that PEP projects could span all business units of Pliva and should not just be confined to GPS manufacturing operations.

Managers from Pharma Chemicals and Local Product Supply who had involvement in PEP emphasised the advantage of a data-driven approach leading to sustainable results. Participants from Regulatory Affairs, Business Development, Medical Marketing Services, Finance and HR shared the potential benefits of extending PEP in their business units but also expressed concerns regarding possible extra workloads, thus straining their resources.

Results and Learning Points

A product manager asked, 'In a simple sentence how would I say to Pliva senior management that PEP is not another fad?' Ron Basu replied, 'PEP has tangible results to show and further savings and improvements will be done by Pliva teams in a Pliva way'. Zeljko Brebric added, 'We also aim to build a team of self sufficient competent managers based upon BB and GB education'.

Leadership workshops or forums were instrumental in gaining the acceptance of the change programme by members of both senior management and middle management.

Source: PEP News, No. 5, January 2006, published by Pliva Group, Zagreb.

Case Study 9: Lean Sigma Changes Abbott's Culture

This case study is an outcome of an MBA dissertation at Henley Management College.

ADC UK is a division of Abbott Inc., which manufactures and distributes medical appliances, in-vitro diagnostic devices and accessories. ADC was formed in April 2004 after a merger between Abbott Laboratories and MediSense Ltd, to become the third largest provider of blood glucose monitoring equipment in the world. Abbot Inc. has operated a succession of policies to improve its financial position. This has culminated in a Lean Sigma implementation in ADC UK from 2006.

A Master (Six Sigma) Black Belt was employed to lead the Lean Sigma initiative, called the Business Excellence Programme. This project was run in parallel with the closure of the original Abingdon site and consolidation to a purpose-built new facility at Witney in Oxfordshire. All of the qualified Black Belts within the Business Excellence team ran approximately four projects at any given time, Green Belt training and mentoring. There had been no new Black Belts trained since the initial education programme. Plans were in place to start a Green Belt upgrading programme during 2008.

A single-site case study approach was utilised to determine the success of this implementation and to ascertain whether the change process led to significant permanent cultural shifts. The effect of governmental regulations on the implementation was also considered. Most previous Lean Sigma initiatives have occurred in unregulated industries.

Data was collected by focused interviews with relevant local management and Lean Sigma trained employees. This information was combined with previously unanalysed statistics, and an internal white paper which correlated quality and regulatory issues with project initiatives at ADCUK.

Results and Learning Points

Analysis demonstrated that the implementation of Lean Sigma into ADCUK had not progressed at the expected rate. These results overall indicated that there were issues in several areas:

- Regulatory requirements.
- Current IT strategy and knowledge management.
- Lean methodology use but less Six Sigma implementation.
- Cultural change.
- Parallel initiatives.

Governmental regulatory issues ensured that this kind of implementation was slower and more complex than the textbook examples for a standard manufacturing company. Projects were shown to take approximately twice as long as in a non-regulated industry. The perceived slow implementation of Lean Sigma was creating some dissatisfaction.

It is recognised that issues in the regulatory environment and the many information streams are significant to the pharmaceutical world. Knowledge management strategy is crucial to a successful implementation. Fragmented IT systems are a hindrance explaining why Lean tools are used in preference to Six Sigma, as they rely less heavily on analysis of quantitative data.

Cultural change had been slow but appeared to be lasting. Furthermore, progress had been made with other initiatives including Kaizen across the site, providing added impetus. The Lean Sigma cultural change programme relies on the site management team to drive a true lasting cultural change, owned by all the employees. One significant aspect of cultural change is the inclusion of suppliers in training and delivering projects. At the time of writing this report, there were 65 Green Belt projects of which 22 projects were led by suppliers.

Source: Sandra E. Goodwin, ADC UK, 2008.

Case Study 10: Visual Six Sigma for a Call Centre Problem

This particular scenario relates to the handling of customer queries via an IT call centre. This call centre does not operate the traditional answering service to complaints, but the details of each call and issue are logged and then passed on to a service engineer capable of solving the problem. The first call is taken by a service engineer who becomes responsible for finding a solution.

Benchmark analysis indicated that the call centre had lower customer satisfaction ratings than best-in-class competitors. Customer satisfaction is the top driver of product revenues, and it is estimated that 8% revenue growth is possible if the company matches the call centre performance of the 'best in class'.

The performance goals to match best-in-class performance are:

- Time to answer should be no more than two minutes.
- 65% of calls must be solved in one iteration, with a maximum service time of 1.5 hours.
- 25% of all calls must be solved with no more than two iterations, with a maximum total service time of five hours.
- 99% of all calls must be solved with no more than three iterations, with a maximum total service time of 15 hours.

Data analysis showed that it took more than two minutes to answer 93.7% of calls, and very few problems were solved within the required duration, irrespective of the number of times a customer contacted the call centre before the problem was solved.

A team was commissioned to investigate the call centre process and dramatically improve process capability. The team applied the lean data analysis process and used cause-and-effect analysis to frame the problem with regard to identification of the potential causes of excessive time to answer and time to solve. The group circled the process inputs in red and these were prioritised as the most likely causes, and were readily available in the call centre database.

Further data collection and analysis by histograms indicated that the only time calls were answered in less than two minutes was on Tuesday and Wednesday afternoons when two staff were on duty. Furthermore, calls were never answered in less than two minutes when three staff were on duty.

This prompted the team to determine what was different about the operation of the call centre when three staff were on duty. Further analysis revealed that only three staff worked on Friday mornings, despite the fact that this was one of the busiest periods.

The distribution of time to answer when there were three staff on duty had a mean of approximately 2.6 minutes, with a range from 2 to 3.2 minutes. Thus it appeared that on some occasions, the staff loading of the call centre had been adjusted to compensate for higher call volumes but that this loading was not optimised to achieve the new performance goals. The impact of increasing the number of staff on duty by one was to reduce the mean time to answer by roughly one minute.

The distribution of time to solve has three clusters of data points: the first, with a mean of around two hours, is associated with calls solved in the first cycle; the second, with a mean of around seven hours, relates to calls solved in the second cycle; and the third, with a mean of around 24 hours, regards those calls solved in the third cycle. Thus, a separate visual analysis was performed for each of the three subgroups defined by the number of cycles.

Figure 11.2 shows simple histograms of each variable when the number of cycles is one. Those calls with a time to solve of less than 1.5 hours are identified with darker shading. This indicates that 'staff experience' followed by 'type of problem' were the top drivers of variation in time to solve.

Results and Learning Points

This example clearly illustrates how a visual approach really facilitates a rapid exploration of the data to quickly drive focus onto the root causes. The modelling and simulation capability then builds on that to allow very rapid testing of 'what if' scenarios and the investigation of different improvement options.

Source: Visual Six Sigma by Ian Cox (www.jmp.com).

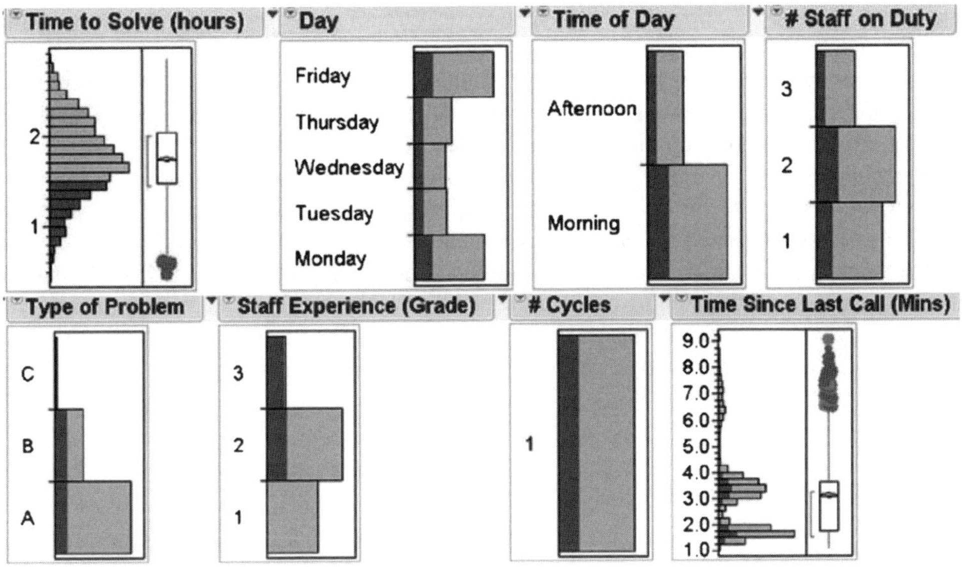

Figure 11.2 Visual exploration of relationships with time to solve when number of cycles = 1

Case Study 11: Empowerment of Middle Management Delivers Results at Novozymes

Novozymes is the world leader in industrial enzymes and biopharmaceutical ingredients. Its products are used in numerous industries such as biofuels, detergents, food and animal feed. Novozymes has more than 40% of the world market for industrial enzymes and is more than twice the size of its nearest competitor. In 2009, the company achieved revenue of DKK 8.5 billion based on a large portfolio of products and services world-wide.

During the period from 2001 to 2005, the yield of major products increased by only 5% as compared to the business plan seeking a 50% increase. On top of that, the members of the Global Optimisation Network (GON) found that the quality of the project plans was poor. This in turn often led to discussion after the project because its scope and objectives were not well defined. The GON member from quality management recommended Six Sigma as the approach to take the improvement culture of Novozymes to the next level.

When presented with the idea of applying Six Sigma as a mandatory approach for improving productivity, Novozymes' top management explicitly stated that they would not support the implementation of Six Sigma because they did not see the instigation of a mandatory methodology as compatible with the Novozymes culture. Novozymes' top management was not only concerned about the cultural fit but just as much whether religious rigidity would destroy innovation and kill all initiative because of bureaucratic project models.

Despite the rejection of turning Six Sigma into a corporate concept, middle management within the quality organisation started promoting Six Sigma and at the first wave approximately 120 employees throughout the organisation received training as either Green or Black Belts. The courses were popular and word-of-mouth became enough to spread the interest of voluntarily participating in Six Sigma training. This indicated that there was a perceived value to Six Sigma even though the worth of implementing Six Sigma in Novozymes had not been evaluated or documented. The first Six Sigma training started in 2004 and evolved in 2007 into Lean Six Sigma, where a part of the statistical training was left out and replaced with fundamental Lean training.

However, it was evident that Novozymes' middle management believed in the value of Lean Six Sigma. This was supported by the fact that more and more employees were voluntarily joining the training programme. Despite this, the support or commitment of those at top management level was not clear. It appeared that they gave some tacit support and wanted to be convinced by tangible results. Novozymes' top management requested a study of the effect of the ongoing Lean Six Sigma effort. They thought it was important to know if the value was only assumed or if it could actually be quantified in any way that could justify the resources used for maintaining the training programme and the continuation of the implementation.

The study was completed as an MBA dissertation at Henley Business School in 2009. The research was conducted as a single case study which included semi-structured interviews with members of top management, Black Belts and Green Belts.

Results and Learning Points

As a consequence of the study, Novozymes has now established a Lean Six Sigma strategy with top management commitment and a cross-functional team that co-ordinates activities. All production managers were trained as Green Belts and the cross-functional group was working on how a Black Belt programme should look at Novozymes.

It can be argued that top management did not commit to the initiative because they believed that dictating the means to achieve a goal was not compatible with Novozymes' culture. The implementation of Lean Six Sigma *was*, however, supported by the culture of Novozymes with a long tradition for working with continuous improvements. Employees were empowered to find their own means to meet the very clear objectives and goals that were cascaded down through all functions. This empowerment was found to be the reason why top management on the other hand would not stop the initiative and explains why middle management carried on with the deployment despite the absence of commitment from top management.

Source: Soren Carlsen, Novozymes A/S (Carlsen, 2009).

11.3 CASE EXAMPLES FOR SERVICES, SMEs, PROJECTS AND GREEN THINKING

The case examples in this section cover organisations other than large manufacturing companies. The examples for services, SMEs, projects and Green thinking underline the FIT SIGMA tenet of 'fitness for the purpose'.

Case Study 12: Kaizen Kept Production in Scotland for Rosti

Rosti Technical Plastics is a medium-sized company with 47 injection moulding machines and a range of ancillary processes including spray painting and ultrasonic welding. The production facilities are situated at Larkhall in Scotland.

Like many other manufacturing industries in the UK and other Western European countries, the viability of Rosti's Larkhall facility is under constant threat from low-cost manufacturing areas such as China and Eastern Europe. In order to remain competitive and secure the future of the Larkhall site, the company was required to reduce the proportion of labour content which went into their products. One such product was a cash cassette assembly of some 37 parts.

With the help of a local consulting firm, the senior management team of Rosti carried out a Kaizen Event against a background of the introduction of the euro note as currency in the EC. The demand for the cash cassette was anticipated to increase twofold. A team of 14 people, including a guest member from the Scottish Executive, was brought together for this exercise. It contained four operators and a number of support people such as internal suppliers and engineers.

The week of the Kaizen Event was the first time that any members of the team had been involved in such a close and intense workgroup. The strength and power of such a team being focused on a single area or problem for a full week was highly productive. The involvement of all the clusters of people in the process made it easy to gain acceptance of what was actually a very different manufacturing process. Many of the operators said they had never before been involved in deciding how their procedures should be set up and run, and they were delighted at this opportunity to participate. The overall feeling of the group at the end of the week was one of exhaustion mixed with exhilaration at having achieved so much in so short a space of time.

Results and Learning Points

The Kaizen Event is a recommended process for FIT SIGMA, especially for small and medium-sized enterprises. This exercise delivered remarkable results, as shown below:

	Goal	Actual
Productivity	50%	67%
Reduce space per cassette	30%	57%
Reduce work in progress	50%	83%
Improve purchased part quality	N/A	90%

Rosti moved on from this success to carry out several more events (including one in Poland) and they are now self-sufficient at running Kaizen Event programmes. Thirty months later the product (which was the subject of the case study) is still being manufactured in Larkhall (not China), thanks to the efforts of the Kaizen team.

Source: Alex Morton, Carrick Lean Consultancy, Scotland, 2010.

Case Study 13: Six Sigma Cuts Network Rail's West Coast Main Line bill

Britain's West Coast Route Modernisation (WCRM) upgrade project was at one point expected to cost more than £10 billion – five times the original forecast. Skilled management and planning, though, brought this down to a more realistic £8 billion, with projects in the programme from 2002 frequently meeting both budget and deadline.

'Six Sigma' methodology has been used successfully by manufacturers for many years and, amongst them, GE and Motorola have been at the forefront of its application. In recent years, many other product manufacturers and service providers have followed their lead for quality improvements, cost reduction and customer satisfaction.

The DMAIC methodology of Six Sigma was relatively new to Network Rail, the sponsor of the WCRM project. However the experience of Bechtel, a major contractor of the project, validated the fact that it can produce quality improvements and generate savings by reducing the cost and volume of rework.

The WCRM is a long-planned upgrade of the 724-km route between London and Glasgow, via Birmingham and Manchester. It is among Europe's busiest mixed-use railways, carrying more than 2000 trains per day. It has 2670 track-km and 13 major junctions.

Line speed was previously limited to 177km/h, but under WCRM it is being raised to 200km/h. The first phase linking London with Manchester was completed in autumn 2004, while the second phase, to Glasgow, was completed by the end of 2006.

Large-scale construction projects such as WCRM have many special characteristics. Some elements constituted new construction without rail traffic operations (blockade approach), while other parts of it were major upgrades under partial line closure. The project generated large quantities of data on various construction activities with common elements of processes. These actions included: design approval process, field reporting procedure, material acquisition and distribution, geometry measurements of constructed track, production tamping and welding processes. All of these courses of action had to comply with Network Rail's (NR) project and railway standards for quality and output of work. These are measurable processes to which Six Sigma can be readily applied.

Recent experience on WCRM suggests that Six Sigma-based process improvement plans work and successfully deliver clear financial benefits for both project and client.

Bechtel brought the approach to WCRM. Both Bechtel and NR each made significant corporate commitments in training project staff to achieve the gains that were theoretically and practically possible by using Six Sigma. Contractors and sub-contractors were also encouraged to participate in PIPs and were delivering better results.

Results and Learning Points

Six Sigma is not a short-term activity. Rather, it requires the serious specialised training of staff and long-term resource commitment, which will definitely pay off in short-term process improvement plans as well as long-term operations.

Railway construction companies can achieve better profitability by drastically reducing the cost of rework, and by delivering the required quality first time. Six Sigma improves service reliability and safety, and ultimately enhances the client's confidence and enriches relations between client and contractor.

The Six Sigma process at Network Rail can be summed up as: identifying the process or activity that is not meeting targets; creating a process improvement team that directly affects the quality and quantity of infrastructure works; and assigning a champion who has a strong interest in the process and who can unblock any difficulties in the implementation plan.

Source: International Railway Journal, June 2005.

Case Study 14: Six Sigma in Radioactive Waste Management Projects

Bechtel was the first major engineering and construction company to adopt Six Sigma, a data-driven approach to improving efficiency and quality in major projects. Here is an example of such application.

The Hanford site in Washington State was established in 1943 to produce plutonium. During Hanford's years of defence production, significant amounts of waste were created and nearly 54 million gallons of high-level radioactive and chemical excess remained in underground storage tanks. In 2001, Bechtel Hanford formed a team called Environmental Restoration Contractor (ERC) with pre-selected contractors to improve the waste management process. Six Sigma methodology (DMAIC) was used to reduce redundancies and improve efficiency in this waste management process. Creating an efficient, cost-effective waste acceptance process entailed integrating dozens of waste-generating activities under five different projects. Each such scheme operated with different goals and regulatory restrictions. More regulatory issues resulted from the low-volume, higher-risk activities.

In addition to deploying the training programme for Black Belts and Green Belts and the application of DMAIC in five different projects, approaches as appropriate to these five projects were adopted. Their purpose was to overcome challenges in four specific areas as follows:

- Quantifying the problem. Efforts were made to utilise the data available in a manner that fitted the task rather than to generate new information. This approach conforms to FIT SIGMA.
- Communication. The process owners from five projects were involved as key members of the process improvement team, encouraging their active participation.
- Application. It was impractical to try and adapt a set of tools derived from the manufacturing world to a project environment. Therefore, the process was adapted to resemble a manufacturing industry and this allowed the acceptance of the appropriate toolkit. This is also in line with FIT SIGMA.
- Differing customer needs. To overcome this challenge, a secondary metric was designed specifically for each project to ensure that the improvements were not having a negative impact on the safety aspect of radioactive waste management.

Results and Learning Points

The Bechtel Hanford-led ERC team worked together with the regional US Department of Energy office to map out failure modes and effect analysis. The application of Six Sigma helped develop an improved process for regulatory compliance and decreased the amount of time needed to process waste.

The process improvements included the development of a complete process flow, the streamlining of waste management procedures and the implementation of waste management training courses at a general and project-specific level. The overall impact of implementing Six Sigma enhanced and sustained the waste management efficiencies of the US Department of Energy.

Source: Waste Management Conference, Tucson, Arizona, 2002.

Case Study 15: Lean Sigma in a Major UK Insurance Company

A major UK insurance company found itself at a crossroads. Plans to grow its market share by 150% were threatened by market changes and economic downturns. Major efforts to streamline internal processes failed. Therefore, the insurer decided to initiate a Lean Sigma-based programme starting with an initial assessment called the 'diagnostic X-ray'.

The approach of the diagnostic X-ray comprised three main activities:

- enterprise value stream mapping;
- benchmarking;
- prioritising.

The move taken by the X-ray team was to develop a value stream map of the organisation's processes and the costs associated with them. The team started by creating a business process fact base looking at all procedures from account set-up to claims handling, gathering data on 189 activities. The analysis showed that there were seven major steps in the value chain. By breaking each stage into the time spent and staffing level, the team found that the company was spending the same amount of time and money on every claim.

The next action was to establish valid benchmarks, both internal and external. The internal benchmarks were created by comparing how various business units carried out similar processes. The team found costly differences. For example, the placement process took twice as long in one business unit as it did in another. Benchmarking also helped the team to track its activity-based costing and evaluate risks.

In the final phase of the X-ray, the team decided which problem to pursue and in which order. This meant plotting the value of solutions against how difficult they would be to implement. With their priorities firmly established, the Black Belts embarked on the Lean Sigma projects.

Results and Learning Points

The diagnostic X-ray approach clearly conforms to the initial assessment of 'fitness for the purpose' in FIT SIGMA and helps to align the resources to business objectives.

There are also other learning points conducive to favourable business results. First, the company set out to reprise some policies supported by activity-based costing. Within the first year the re-pricing alone earned the insurer an additional 10% of profits. Second, Black Belts worked on a series of targeted organisational changes, boosting profits by a further 10%. Third, over the next 30 months the company invested in technology upgrades justified by focused process improvements. Ultimately, IT investments helped to deliver the final 30% profit increase.

Source: Bain & Co. (www.bain.com).

Case Study 16: Environmental Management Dashboard Slashes a Bank's Carbon Footprint

In 2007, a major UK bank, employing 140,000 people across 30 countries, reduced its carbon footprint by 210,000 tonnes of CO_2, which was 36% of its total footprint.

The tool responsible for this achievement was the Environmental Management Dashboard (the footprint tracker) by Building Sustainability Ltd (see Figure 11.3). The dashboard provided a flexible means to visualise the total environmental load with special focus on energy consumption, energy savings and energy generations by people, equipment, buildings and complete property portfolios. By measuring and presenting the actual cost of energy, energy consumption and CO_2 emission (carbon footprint), the occupants, engineers and managers were motivated to introduce energy-efficient solutions.

The bank was using the dashboard system in a controlled experiment to measure the impact of carbon emissions on renewable energy sources and on enabling employees to control their

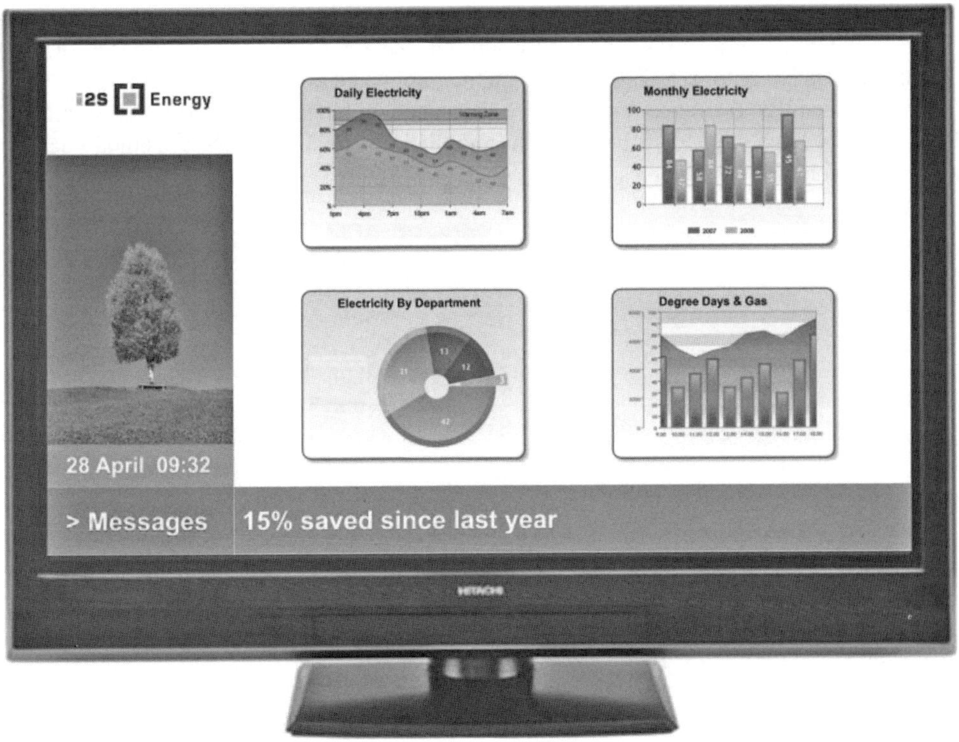

Figure 11.3 Environmental Management Dashboard

individual energy usage. The innovative new approaches employed included a digital lighting control system, roof-top photovoltaic cells and a solar thermal panel for water heating.

Results and Learning Points

The high profile and intuitive ease of use of the carbon dashboard enabled the bank to demonstrate its environmental credentials and commitment to progressive technologies, as well as its corporate social responsibility to employees and customers alike. During the period of the 'credit crunch', all banks were unpopular to say the least, and this initiative was considered as an acceptable 'bonus' by the bank concerned. The visual and user-friendly system also validated the saying 'you cannot manage what you cannot measure'.

The above case example conforms to FIT SIGMA in Green thinking.

Source: Building Sustainability Limited (www.buildingsustainability.net).

11.4 SUMMARY

In this chapter, case studies are included to offer a practical insight to their applications in both manufacturing and non-manufacturing organisations. We have chosen examples of both company-wide programmes and solutions to problems so that both the application of tools and techniques and the holistic approach of FIT SIGMA implementation can be illustrated. The case examples have been simplified to present the key learnings and approach of each example that may conform to FIT SIGMA methodology and objectives.

It is intended that the reader may attain a sharpened understanding of how Six Sigma and Lean projects have worked in practice, in compliance with FIT SIGMA methodology, as appropriate for each case.

Appendix I
Questions and Exercises

1. In the context of your organisation and the status of quality improvement initiatives, examine the steps of implementation described in Chapter 10 and develop an implementation strategy to move towards total quality management, Six Sigma or FIT SIGMA as appropriate.
2. A large multinational FMCG manufacturing organisation has a turnover of £12 billion p.a. and operates in 32 countries with approximately 60,000 employees. Both the growth and the profitability of the company have been declining over the last five years. Some of the subsidiary companies have carried out isolated cost-effectiveness programmes under different banners, e.g. TQM, TPM, BPR and JIT. What would be your recommended continuous improvement strategy for the multinational company?
3. You have been asked as a management consultant, with particular expertise in Six Sigma, to develop a corporate strategy for rolling out operational excellence programmes to all subsidiary companies. Outline a plan to include:
   - Objectives and scope.
   - Corporate imperatives.
   - Methodology, tools and techniques.
   - Estimated costs and benefits.
   - Implementation steps.
   - Rollout time plan.
4. Your organisation has been successful in the implementation of a quality programme based on Six Sigma principles. Write a case study of the programme by taking into account:
   - Organisation overview (whether service or manufacturing).
   - Drivers for change.
   - How DMAIC methodology has been used.
   - What kind of tools and techniques have been used and why.
   - Costs and benefits.
   - Learning points.
5. Risk AB in Sweden is an insurance broker and risk consultancy firm and a wholly owned subsidiary of a stock-tested US group. The company has a turnover of SK 150 million and employs 100 people. Having read the success stories of GE Capital, the CEO of the company is keen to consider an appropriate service excellence programme based on Six Sigma principles. As an internal consultant, how would you develop an action plan?
6. Medicare plc is a world-class pharmaceutical company of 18,000 employees and £15 billion turnover. The company has successfully implemented a Lean Sigma programme. Develop a strategy for the company to continuously improve and sustain the high performance levels.
7. Infotec is a world leader in computer hardware. Over the last decade the software service side of the business has experienced the most significant growth, now accounting for 40%

of the total revenue. The company experimented with a TQM programme in the early 1990s but it was abandoned after two years. The newly appointed CEO wishes to initiate an EFQM-driven quality management programme in the EMEA region of the company. Develop an implementation strategy for the CEO.

8. What are the features and philosophy common to both TQM and Six Sigma?
   Explain the new features, if any, in a Six Sigma programme.
   What are the additional features in Lean Sigma and FIT SIGMA?
9. What are Deming's '14 criteria for quality improvement'?
   Explain Deming's 'quality wheel'.
   With appropriate reference to the 'quality guru', explain the following:
   - 'Fitness for the purpose'.
   - 'Quality is free'.
   - 'Moment of truth'.
   - 'Fishbone diagram'.
   - SMED.
   - FIT SIGMA.
10. Explain why tools and techniques are essential in the implementation of quality and operational excellence programmes.
    Recommend appropriate DMAIC methodology for a FIT SIGMA programme to be applied in:
    - A large multinational bank.
    - A manufacturing company employing 200 people.
    - A mini-market employing 20 people.
11. Describe the main driving and resisting forces in the implementation of a change programme.
    What are the critical success factors related to the application of tools and techniques?
12. In your organisation, which processes have the highest priority for improvement?
    Which tools would you use to come to this conclusion?
13. In your local hospital, even after their admission to a specific ward, patients have to wait for their beds. Use the cause and effect diagram to show the possible causes for the delay in assigning beds.
14. You are going to buy 10 different items from your local supermarket. The major activities involved would include parking the car, collecting your trolley, picking items you wish to buy, paying at the check-out, returning to your car and finally unloading purchases into it. Develop a flow process chart by taking into account detailed elements of operations and identifying the non-value-added activities.
15. What are the distinctive features of:
    - value stream mapping;
    - process mapping;
    - flow diagram;
    - flow process chart.

    Give examples of the areas of application of these tools. Explain how these tools can be applied for both service and manufacturing operations.
16. Discuss the relationship between
    (a) brainstorming, nominal group technique and mind mapping;
    (b) a run chart, scatter diagram and regression analysis;
    (c) an inter-relationship diagram and affinity diagram;

(d) control charts and process capability measurement;
(e) OEE and SMED;
(f) the five Ss, the five whys and mistake-proofing;
(g) a Gantt chart, activity network diagram and milestone tracker diagram.

17. As a management consultant with expertise in operational excellence, recommend appropriate tools and techniques for the following programmes:
    (a) FIT SIGMA in a chain of hotels.
    (b) TQM in a mail-order firm.
    (c) Lean Enterprise in the order processing department of a large telecommunication company.
    (d) Cost improvement in a road construction company.
    (e) New process development in a research organisation.
    (f) Improving packaging effectiveness of a FMCG manufacturing company.

Appendix II
Introduction to Basic Statistics

STATISTICS

Statistics is the art and science of using numerical facts and figures. In Wikipedia, statistics is defined as a mathematical science pertaining to the collection, analysis, interpretation or explanation, and presentation of data.

There are three kinds of statistics:

- descriptive statistics;
- inferential statistics;
- causal modelling.

Descriptive statistics deals primarily with the description and interpretation of data by figures, graphs and charts.

Inferential statistics is the science of making decisions in the face of uncertainty by using techniques such as sampling and probability.

Causal modelling is part of inferential statistics. It is aimed at advancing reasonable hypotheses about underlying causal relationships between the dependent and independent variables.

In Six Sigma projects the most frequently used statistics is descriptive statistics, and the most useful distribution is the normal distribution.

Descriptive Statistics

- Data distribution: normal distribution, other distributions.
- Measures of central tendency: mean, median, mode.
- Measures of dispersion: standard deviation, variance and range.

Normal Distribution

Most data tends to follow the **normal distribution**, or bell-shaped curve. One of the key properties of the normal distribution is the **relationship** between the **shape of the curve and the standard deviation**.

99.73% of the area of the normal distribution is contained between –3 sigma and +3 sigma from the mean. Another way of expressing this is that 0.27% of the data falls outside 3 standard deviations of the mean.

The population is the total number in the study, e.g. census, and a sample is a small subset of the population. Mean and standard deviation are expressed by Greek letters (μ, σ) in the population and by Roman letters (X, s).

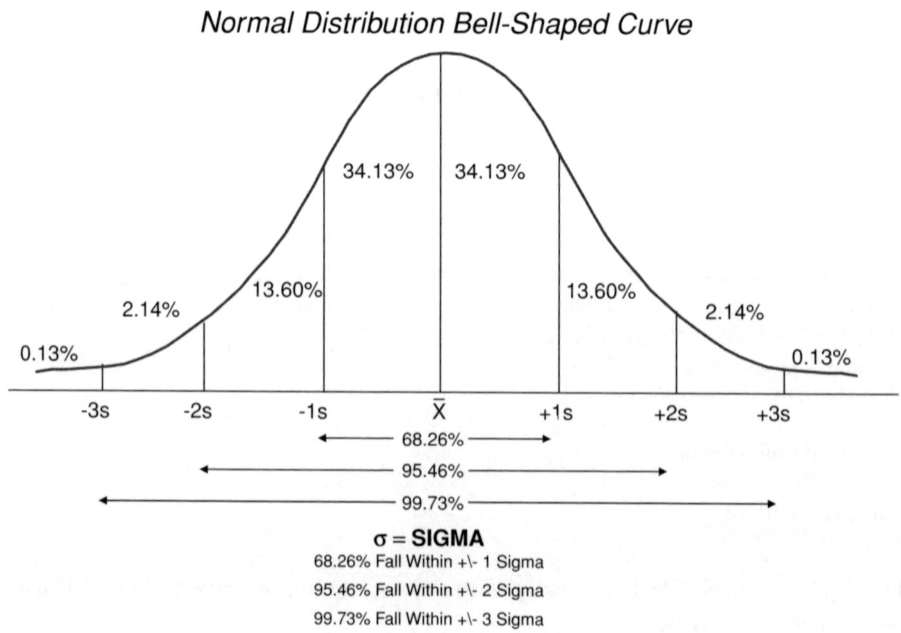

Measures of Central Tendency

Measures of central tendency are determined by mean, median and mode.

Mean is the arithmetic average of a set of data. It is a measure of central tendency, not a measure of variation. However, it is required to calculate some of the statistical measures of variation.

$$\overline{x} = \frac{\sum_{i=1}^{n} x_i}{n-1}$$

where: $\bar{x}$ = sample mean
x_i = data point i
n = group sample size

Median is the central or middle data point, e.g. in the series 32, 33, 34, 34, 35, 37, 37, 39, 41 the median is 35.

Mode is the highest data point, e.g. in the series 32, 33, 34, 34, 35, 37, 37, 39, 41 the mode is 41.

Measures of Dispersion

The measure of dispersion is determined by standard deviation, variance and range, and the shape of the distribution is measured by skewness and kurtosis.

Range is the simplest measure of dispersion. It defines the spread of the data and is the distance between the largest and the smallest values of a sample frequency distribution.

Variance is the average of the square values of the deviations from the mean.

Standard deviation is defined as follows:

$$s = \sqrt{\sum_{i=1}^{n} \frac{(x_i - \bar{x})^2}{n - 1}}$$

It is the square root of variance and in a more intuitive definition, think of it as the 'average distance from each data point to the mean'.

Skewness and Kurtosis

Skewness measures the departure from a symmetrical distribution. When the tail of the distribution stretches to the left (smaller values) it is negatively skewed and when it stretches to the right it is positively skewed.

Kurtosis is a measure of distribution 'peakness' or flatness. A curve is too peaked when the kurtosis exceeds +3 and is too flat when the kurtosis is less than –3.

Sources of Variation

There are two types or sources of variation, as shown in the following table:

Type	Definition	Characteristics
Common cause	*No way to remove*	***Always present***
	Influenced by one of the 6 Ms	*Expected* *Normal* *Random*
Special cause	*Can be removed*	***Not always present***
	Influence of the 6Ms	*Unexpected* *Not normal* *Not random*

To distinguish between Common cause and Special causes variation, use display tools that study variation over time such as Run Charts and Control Charts.

Process Capability

Process capability refers to the ability of a process to produce a product that meets given specifications.

It is measured by one or a combination of four indices:

DPM	defects per million
σ level	number of standard deviations from the centre
C_p	potential process index
C_{pk}	process capability index

$$\sigma\text{level} = \text{minimum } (\text{USL} - \mu)/\sigma, (\mu - \text{LSL})/\sigma$$

where: USL = upper specification level
LSL = lower specification level
σ = standard deviation
μ = mean
C_p = process capability index
= minimum $(USL - \mu)/3\sigma, (\mu - LSL)/3\sigma$
= σ level/3

$$C_p = \frac{|USL - LSL|}{6\sigma}$$

Process Sigma

Process sigma is an expression of process yield based on DPMO (defects per million opportunities) to define variation relative to customer specification. A higher value process sigma is an indication of a lower variation relative to specification.

Historical data indicates a change of sigma value by 1.5 when long-term and short-term (actual) process sigma values are compared. The long-term values are higher by 1.5, e.g. 6 sigma is long-term process sigma for 3.4 DPMO.

A sample process sigma calculation:
Given 7 defects (D) for 100 units (N) processed with 2 defect opportunities (O) per unit.
Defects per opportunity, DPO = $D/(N \times O) = 7/(100 \times 2) = 0.035$
Yield = $(1 - \text{DPO}) \times 100 = 96.5\%$
From the yield conversion table (see Appendix III):
long-term process sigma = 3.3
actual process sigma = 3.3 – 1.5

Inferential Statistics

Inferential statistics helps us make judgements about the population from a sample. A sample is a small subset of the total number in a population. Sample statistics are summary values of samples and are calculated using all the values of the sample. Population parameters are summary values of the population but they are not known. That is why we use sample statistics to infer population parameters.

$$\text{Sample size SS} = (\text{DC} \times \text{V/DP})^2$$

where: DC = degree of confidence = number of standard errors for the degree of confidence
V = variability = standard deviation of the population
DP = desired precision = acceptable difference between the sample estimate and the population value.

A **null hypothesis** refers to a population parameter and not a sample statistic. Based on the sample data the researcher can accept the null hypothesis or accept the alternative hypothesis.

The ***t-test*** compares the actual difference between two means in relation to the variation in the data (expressed as the standard deviation of the difference between the means). It is applied when sample sizes are small enough (less than 30) using an assumption of normality. A test of the null hypothesis is that the means of two normally distributed populations are equal.

Analysis of variance (**ANOVA**) is used to assess the statistical differences between the means of two or more groups of a population. ANOVA can also examine research problems that involve several independent variables.

The F-test assesses the differences between the group means when we use ANOVA.

F = variance between groups/variance within groups. Larger F-ratios indicate significant differences between the groups and most likely the null hypothesis will be rejected.

Causal Modelling

Many business objectives are concerned with the relationship between two or more variables. Variables are linked together if they exhibit **co-variation**; that is, when one variable consistently changes relative to the other variable. A **coefficient of correlation** is used to assess this linkage. Large coefficients indicate high co-variation and a strong relationship, and vice versa.

In causal modelling studies it is assumed that some variables cause or effect other variables, and the aim is to confirm or refute the hypothesised relationships and estimate their nature and robustness. The value of a model is to add clarity and understanding to an otherwise complex situation, to understand the relationship between variables and to enable a theory to be tested. The analysis of causal models is carried out by structural equation modelling (SEM)-based techniques.

The advantage that SEM-based techniques have over first-generation techniques (such as multiple regression analysis) is the greater flexibility that a researcher has for the interplay of theory and data. SEM enables the researcher to construct immeasurable variables measured by indictors (also called observed variables) as well as measurement errors. Hence it overcomes the limitations of first-generation techniques such as multiple regression.

SPSS

The data analysis and statistical techniques presented in this appendix are all available in the popular software package SPSS (www.spss.com).

Appendix III
Yield Conversion Table

These are estimates of long-term sigma values. Subtract 1.5 from these values to obtain actual sigma values.

Sigma	DPMO	Yield %	Sigma	DPMO	Yield %
6.0	3.4	99.99966	3.0	66897	93.3
5.9	5.4	99.99946	2.9	80757	91.9
5.8	8.5	99.99915	2.6	90801	90.3
5.7	13	99.99866	2.7	155070	88.5
5.6	21	99.9979	2.6	135666	86.4
5.5	32	99.9968	2.5	158655	84.1
5.4	48	99.9952	2.4	184060	81.6
5.3	72	99.9928	2.3	211855	78.8
5.2	108	99.9892	2.2	241964	75.8
5.2	159	99.984	2.1	374253	72.6
5.0	233	99.977	2.0	308538	69.1
4.9	337	99.966	1.9	344578	65.5
4.8	483	99.952	1.8	382089	61.8
4.7	687	99.931	1.7	420740	57.9
4.6	968	99.90	1.6	460172	54.0
4.5	1350	99.87	1.5	500000	50.0
4.4	1866	99.81	1.4	539828	46.0
4.3	2555	99.74	1.3	579260	42.1
4.2	3467	99.65	1.2	617911	38.2
4.1	661	99.53	1.1	665422	34.5
4.0	6210	99.38	1.0	691482	30.9
3.9	198	99.18	0.9	725747	27.4
3.8	10724	98.9	0.8	758036	24.2
3.7	13903	98.6	0.7	788145	21.2
3.6	17864	98.2	0.6	815940	18.4
3.5	22750	97.7	0.5	841345	15.9
3.4	28716	97.1	0.4	864334	13.6
3.3	35930	96.4	0.3	884930	11.5
3.2	44565	95.5	0.2	903199	9.7
3.1	54799	94.5	0.1	919243	8.1

References

ACE (2008) Allied Consultants Europe, *Operational and Lean Management survey 2008*, www.implement.dk.

Achanga, P., Shehab, E., Roy, P. and Nedler, G. (2006) 'Critical success factors for Lean implementation within SMEs', *Journal of Manufacturing Technology Management*, **17**(4), 460–471.

Ala-Risku, T. and Karkkainen, M. (2006) 'Material delivery problems in construction: a possible solution', *International Journal of Production Economics*, **104**(1), 19–29.

Albrecht, K. (1988) *At America's Service*, Dow-Jones Irwin, Homewood, IL.

Antony, J., Kumar, M. and Madu, C.N. (2005) 'Six sigma in small- and medium-sized UK manufacturing enterprises: some empirical observations', *International Journal of Quality & Reliability Management*, **22**(8), 860–874.

Axelrod, R.H. (2001) 'Changing the way we change organizations', *The Journal of Quality and Participation*, **24**(1), 22–27.

Barfod, N. (2007) 'Man behøver ikke være stor for at være lean (You don't have to be big to be lean)'. *Børsen*, **5**(15), 13–14.

Barlett, C. and Ghoshal, S. (1994) *The Changing Role of Top Management*, INSEAD Working Papers, London Business School Library, London.

Bassart, M. and Ritter, D. (1994) *The Memory Jogger*, Goal/QPC, Methuen, Massachusetts.

Basu, R. (2004) 'Six Sigma in operational excellence: role of tools and techniques', *International Journal of Six Sigma and Competitive Advantage*, **1**(1), 44–64.

Basu, R. (2009) *Implementing Six Sigma and Lean*, Elsevier, Oxford.

Basu, R. (2010) 'In search of project excellence: a systems approach of triangulation in a mixed methodology empirical research', *International Journal of Business Systems Research*, **4**(4), 432–450.

Basu, R. and Wright, J.N. (1998) *Total Manufacturing Solutions*, Butterworth-Heinemann, Oxford.

Basu, R. and Wright, J.N. (2003) *Quality Beyond Six Sigma*, Elsevier, Oxford.

Basu, R. and Wright, J.N. (2005) *Total Operations Solutions*, Butterworth-Heinemann, Oxford.

Basu, R. and Wright, J.N. (2008) *Total Supply Chain Management*, Elsevier, Oxford.

Berr (2008) *Small and Medium Enterprise Statistics for the UK and Regions*, Berr Enterprise Directorate Analytical Unit, London.

Best Practices LLC (2002) *Building Six Sigma Excellence: Case Study of General Electric*, Best Practices LLC, North Carolina.

Black, J.T. and Phillips, D.T. (2010) 'The lean to green evolution', *Industrial Engineer*, **42**(7), 46–51.

Boddy, D. and Macbeth, D. (2000) 'Prescriptions for managing change: a survey of their effects in projects to implement collaboration with organisations', *International Journal of Project Management*, **18**(5), 297–306.

Brenig-Jones, M. (2008) 'Introducing Kai Sigma', *Quality World*, June.

Breyfogle III, F.W. (2008) 'Better fostering innovation: 9 steps that improve Lean Six Sigma', *Business Performance Management*, September.

Carlsen, S. (2009) 'The role of top management in the successful implementation of Lean Six Sigma in a Danish manufacturing company', Unpublished MBA dissertation, Henley Business School.

Carnall, C. (1999) *Managing Change in Organisations*, Prentice-Hall, London.

Corporate Leadership Council (2005) 'Leadership competencies in a Lean environment', Corporate Executive Board, August 2005, Catalog no. CLC13TKO7X, London.

Crosby, P.B. (1984) *Quality Without Tears*, McGraw-Hill, London.

Davies, C.E. (2005) 'South West Region: Six Sigma for SMEs', *Management Services*, Summer, 6–7.

De Carlo, N. (2007) *Lean Six Sigma*, Alpha Books, New York.

Deming, W.E. (1986) *Out of the Crisis*, MIT Centre for Advanced Research, Cambridge, MA.

Deshmukh, S.V. and Lakhe, R.R. (2009) 'An innovation model of Six Sigma for SMEs: the T-DMAIC model', *International Journal of Six Sigma and Competitive Advantage*, **5**(2), 187–203.

Easton, G. and Jarrell, S. (1998) 'The effects of total quality management on corporate performance', *Journal of Business*, **71**(3), 253–307.

Eckes, G. (2001) *Making Six Sigma Last*, John Wiley & Sons, New York.

Elkington, J. (1994) 'Towards the sustainable corporation: win–win–win business strategies for sustainable development', *California Management Review*, **36**(2), 90–100.

Erwin, J. and Douglas, P.C. (2000) 'It is difficult to change company culture', *Journal of Supervision*, **51**(11).

EU (2007) *Observatory of European SMEs*, European Commission, Flash EB Series # 106.

European Foundation of Quality Management (2003) *The EFQM Excellence Model*, Brussels.

Friedman, M. (1990) *Free to Choose: A Personal Statement*, Harvest Books, Fort Washington, PA.

Gabor, A. (2000) 'He made America think about quality', *Fortune*, **42**(10).

George, M.L. (2002) *Lean Six Sigma: Combining Six Sigma Quality with Lean Production Speed*, McGraw-Hill, New York.

George, M.L., Rowland, D., Price, M. and Maxey, J. (2005) *Lean Sigma Pocket Tool Book*, McGraw-Hill, New York.

Goldratt, E.M. (1992) *The Goal*, North River Press, New York.

Hall, J. (1999) 'Six principles of successful business change management', *Management Science*, **43**(4), 16–18.

Hammer, M. and Champy, J. (1993) *Re-Engineering the Corporation*, Nicholas Brealey Publishing, London.

Harry, M. and Crawford, J.D. (2004) 'Six sigma for the little guy', *Mechanical Engineering*, **126**(11), 8–10.

Hartman, E. (1991) *'How to install TPM in your plant', 8th International Maintenance Conference*, Dallas, 12–14 November.

Ishikawa, K. (1985) *What is Total Quality Control? A Japanese Way*, Translated by D.J. Lu, Prentice-Hall, London.

Kaplan, R.S. and Norton, D.P. (2004) 'Measuring the strategic readiness of intangible assets', *Harvard Business Review*, **82**(2), 52–64.

Knuckey, S., Leung-Wai, J. and Meskill, M. (1999) *Gearing Up: A Study of Best Manufacturing Practices in New Zealand*, New Zealand Ministry of Commerce, Wellington.

Kotabe, M. and Hensen, K. (2000) *Global Marketing Management*, John Wiley & Sons, New York.

Kotler, P. (1999) *Kotler on Marketing*, Free Press, New York.

Kumar, M., Antony, J., Singh, R.K., Tiwari, M.K. and Perry, D. (2006) 'Implementing the Lean Sigma framework in an Indian SME: a case study', *Production Planning and Control*, **17**(4), 407–423.

Ladhar, H. (2007) 'Effective Lean Six Sigma deployment in a global EMS environment', *Circuits Assembly*, March.

Ledolter, J. and Burnhill, C.W. (1999) *Achieving Quality Through Continual Improvement*, John Wiley & Sons, New York.

Lewin, K. (1951) *Field Theory in Social Science: Selected Theoretical Papers*, Harper-Collins, New York.

Lewis, B.R. (1994) in Dale, B.G. (ed.), *Managing Quality*, Prentice-Hall, London.

Liker, J.K. (2004) *The Toyota Way*, McGraw-Hill, New York.

Ling, R.C. and Goddard, W.E. (1988) *Orchestrating Success*, John Wiley & Sons, New York.

Machiavelli, N. (1513) *The Prince*, Translated by R. Luigi, New American Library of World Literature, New York.

Mengel, T. (2008) 'Outcome-based project management education for emerging leaders – a case study of teaching and learning project management', *International Journal of Project Management*, **26**(3), 275–285.

Mintzberg, H. (1996) 'Musings on management', *Harvard Business Review* **July–August**, 61–67.
Moroney, M.J. (1973) *Facts from Figures*, Penguin Books, London.
MSP (2007) *Managing Successful Programme*, 3rd edn, The Office of Government Commerce, London.
Oakland, J.S. (2003) *'QM: Text with Cases*, Butterworth-Heinemann, Oxford.
Obeng, E. (1994) *All Change! The Project Leaders' Secret Handbook*, Pitman, London.
Ohno, T. (1988) *Toyota Production Systems*, Productivity Press, Cambridge, MA.
Pearce, I. (1992) 'Thinking green', *Environmental Management and Health*, **3**(2), 6–12.
Peters, T. and Austin, N. (1986) *A Passion for Excellence*, Fortune, London.
Project Management Institute (2004) *A guide to the project management body of knowledge (PMBOK Guide)*, 3rd edn, Newtown Square.
Pyzdek, T. (2003) *The Six Sigma Handbook*, McGraw-Hill, New York.
Schmidt, S.R., Kiemele, M.J. and Berdine, R.J. (1999) *Knowledge Based Management*, Academy Press, Colorado Springs, CO.
Schonberger, R. (1986) *World Class Manufacturing*, The Free Press, New York.
Shingo, S. (1985) *Non- Stock Production*, Productivity Press, Cambridge, MA.
Shirose, K. (1992) *TPM for Workshop Learners*, Productivity Press, Cambridge, MA.
Skinner, S. (2001) 'Mastering basic tenets of Lean Manufacturing – Five Ss', *Manufacturing News*, **8**(11).
Snee, R.D. and Hoerl, R.W. (2009) 'Turning to service sectors', *Industrial Engineer*, **October**, 37–40.
Stamatis, D.H. (1999) *Six Sigma and Beyond*, St Lucie Press, New York.
Sterman, J., Repenning, N.P. and Kofman, F. (1997) 'Unanticipated side effects of successful quality programmes: exploring a paradox of organizational improvement', *Management Science*, **43**(4), 503–521.
Stern, N. (2006) *Stern Review: The Economics of Climate Change*, HM Treasury, London.
Tuckman, B.W. (1965) 'Development sequence in small groups', *Psychological Bulletin*, **63**, 384–399.
Turner, J.R. (1999) *The Handbook of Project Based Management*, 2nd edn, McGraw-Hill, London.
Turner, J.R., Grude, K.V. and Thurloway, L. (1996) *The Project Managers as Change Agents*, McGraw-Hill, London.
Turner, J.R. and Simister, S.J. (2000) *Gower Handbook of Project Management*, Gower Publishing, Aldershot.
Treacy, M. and Wiersema, F. (1993) 'Customer intimacy and other value discipline', *Harvard Business Review*, **Jan/Feb**, 84–93.
United Nations (1987) *Report of the World Commission on Environment and Development*, UN General Assembly Resolution, 11 December, New York.
Weinstein, L.B., Castellano, J., Petrick, J. and Vokurka, R.J. (2008) 'Integrating Six Sigma concepts in an MBA Quality Management Class', *Journal of Education for Business*, **83**(4), 233–238.
Welch, J. (2001) *Jack: What I have learned leading a great company and great people*, Headline Book Publishing, London.
Wessel, G. and Butcher, P. (2004) 'Six Sigma for small and medium-sized enterprises', *TQM Magazine*, **16**(4), 264–272.
Wild, R. (2002) *Operations Management*, Continuum, London.
Womack, J.P. and Jones, D.T. (1998) *Lean Thinking*, Touchstone Books, London.
Wright, J.N. (1999) *The Management of Service Operations*, Cassell, London.
Wright, J.N. and Basu, R. (2008) 'Project management and Six Sigma: obtaining a fit', *International Journal of Six Sigma and Competitive Advantage*, **4**(1), 81–94.
Yu, J., Williams, E., Ju, M and Yang, Y. (2010) 'Forecasting global generation of obsolete personal computers', *Environmental Science and Technology*, **44**(9), 3232–3237.
Zeithaml, V.A., Prasuraman, A. and Barry, L.L. (1990) *Delivering Service Quality*, Free Press, New York.

Glossary

ABC analysis Based on Pareto analysis, grouping units usually according to the share of annual cost. Units having 80% annual cost are considered in the 'A' classification, units with the bottom 5% share are 'C' items and units with costs in between are in the 'B' category.

Activity based costing Analysing the cost of an operation at each processing step. In addition to measuring direct costs it covers bottlenecks, delays and other time-related activities to highlight areas of inefficiency in an operation.

Activity network diagram A network analysis technique to allow a team to find the most efficient path and realistic schedule of a project by showing graphically the completion time and sequence of each task.

Affinity diagram Used to generate a number of ideas by a team and then organise natural groupings among them to understand the essence of a problem.

Bar chart Also known as a Gantt chart, indicates scheduling activities. Horizontal bars show the various activities, with the length of the bar proportional to the duration of a particular activity.

Benchmarking Rating an organisation's products, processes and performances with other organisations in the same or another business. The objective is to identify the gaps with competitors and the areas for improvement.

Best practice Refers to any organisation that performs as well as or better than the competition in quality, timeliness, flexibility and innovation. Best practice should lead to world-class performance.

Black Belts Experts in Six Sigma methods and tools. Tools include statistical analysis. Black Belts are project leaders for Six Sigma initiatives; they also train other staff members in Six Sigma techniques.

BPR Business process re-engineering has been described as a manifesto for revolution. The approach is similar to taking a clean piece of paper and starting all over by identifying what is really needed to make the mission of the organisation happen.

Brainstorming A free-wheeling group session for generating ideas. Typically a group meeting of about seven people will be presented with a problem. Each member will be encouraged to make suggestions without fear of criticism. One suggestion will lead to another. All suggestions, no matter how seemingly fanciful, are recorded and subsequently analysed. Brainstorming is useful for generating ideas for further detailed analysis.

Carbon footprint The total set of greenhouse gas (GHG) emissions caused by an organisation, event or product. For simplicity of reporting, it is often expressed in terms of the amount of carbon dioxide or its equivalent of other GHGs emitted. A carbon footprint can be measured by undertaking a GHG emissions assessment of an individual, organisation or country.

Cause and effect diagram The cause and effect, fishbone or Ishikawa diagram was developed by Kaoru Ishikawa. The premise is that generally, when a problem occurs, the effect is very obvious; and the temptation is to treat the effect. With the Ishikawa approach the causes of the effect are sought. Once the cause is known and eliminated, the effect will not be seen again. For example, working overtime is an effect, adding extra staff does not remove the cause. The question is what caused the situation that led to overtime being worked.

Check sheet A method of systematically recording data from historical sources or observations as they happen. The patterns and trends can be clearly detected and shown.

Continuous improvement Always looking for ways to improve a process or a product, but not necessarily making radical step changes. If the basic idea is sound then building on it will improve quality. In Japan this is known as Kaizen.

Control chart A tool in statistical process control to monitor the number of defects found in a product or a process over time and study the variation and its source.

COPQ The cost of poor quality is made up of costs arising from internal failures, external failures, appraisal, prevention and lost opportunity costs. In other words, all the costs that arise from non-conformance to a standard. Chapter 3 discusses COPQ in some detail.

CRM Customer relationship management is the development of the database and strategies necessary to have the maximum client relationships in terms of quality, cost, reliability and responsiveness.

CTQs In Six Sigma, critical to quality. This simply means the identification of factors that are critical for the achievement of a level of quality.

Cycle time The elapsed time between two successive operations or the time required to complete an operation.

Digital Six Sigma (DSS) Reported to be a revitalisation of Six Sigma methodology by Motorola through the following improvements:

- Leveraging new digital tools to drive project success.
- Digitising business processes to 'permanently' enforce optimal process compliance.
- Tracking vital processes through digital databases.
- 'Permanently' and proactively eliminating sources of variation that cause defects.

DFSS Design for Six Sigma, see Chapter 9 for detailed discussion. The steps are Define, Measure, Analyse, Design and Validate.

DMAIC The cycle of Define, Measure, Analyse, Improve and Control, see Chapter 9 for detailed discussion.

DOE The process of examining options in the design of a product or service. Controlled changes of input factors are made, and the resulting changes to outputs noted. Losses from poor design include not only direct loss to the company from reworking and scrap, but downtime for the user due to equipment failure, poor performance and unreliability. Poor customer satisfaction will lead to further losses by the company as market share falls.

DPMO Defects per million opportunities. This is the basic measure of Six Sigma. It is the number of defects per unit divided by the number of opportunities for defects multiplied by 1,000,000. This number can be converted into a Sigma value. For example, Six Sigma = 3.4 per million opportunities.

DRP Distribution requirement planning is the planning step in the supply chain to move finished goods from production or stock to the customer.

Earned value management (or earned value analysis) A project control tool for comparing the achieved value of work in progress against the project schedule and budget. It can be performed at the single activity level and by aggregating the results up through the hierarchy or work breakdown structure.

E-business Electronic business is more than the transfer of information using information technology. E-business is a complex mix of processes, applications and organisational structures.

EFQM The European Foundation for Quality Management is derived from the American Malcolm Baldridge Quality award. It is an award for organisations that achieve world-class performance as judged by independent auditors against a checklist. The checklist is detailed and extensive and covers: Leadership, People Management, Policy and Strategy, Partnerships and Resource, Processes, People Satisfaction, Customer Satisfaction, Impact on Society, Business Results

ERP Enterprise resource planning is the extension of MRP(II) systems to the management of complete business functions including Finance and Human Resources.

Failure mode and effect analysis FMEA was developed in the aerospace and defence industries. It is a systematic and analytical quality planning tool to identify, for new products or services at the design stage, what could go wrong during manufacture, or when in use by the customer. It is an iterative process and questions asked/examined are:

1. What is the function?
2. Potential failure modes.
3. The effect of potential failure.
4. Review of current controls.
5. Determination of risk priority (occurrence, detection and severity of failure).
6. Identification of corrective actions to eliminate failure.
7. Monitoring of corrective actions and countermeasures.

First pass yield FPY, also known as RTY, is the ratio of the number of completely defect-free units at the end of a process (without any kind of rework during the process) and the total number of units at the start of a process. The theoretical throughput rate is often regarded as the number of units at the start of the process. RTY/FPY is used as a key performance indicator to measure overall process effectiveness.

Fishbone diagram The fishbone, Ishikawa or cause and effect diagram was developed by Kaoru Ishikawa. The premise is that generally, when a problem occurs, the effect is very obvious; and the temptation is to treat the effect. With the Ishikawa approach the causes of the effect are sought. Once the cause is known and eliminated, the effect will not be seen again. For example, working overtime is an effect, adding extra staff does not remove the cause. The question is what caused the situation that led to overtime being worked.

FIT SIGMA^{TM}, *see also* TQM, Six Sigma and Lean Sigma FIT SIGMA incorporates all the advantages and tools of TQM, Six Sigma and Lean Sigma. The aim is to get an organisation healthy (fit) by using appropriate tools for the size and nature of the business (fitness for the purpose) and to sustain a level of fitness. FIT SIGMA is a holistic approach.

Five Ss These represent a set of Japanese words for excellent house-keeping: Sein, sort; Seiton, set in place; Seiso, shine; Seiketso, standardise; Sitsuke, sustain.

Flow process chart A flow process chart sets out the sequence of the flow of a product or a procedure by recording all the activities in a process. The chart can be used to identify steps in the process, value-adding activities and non-value-adding activities.

FMEA Failure mode and effect analysis was developed in the aerospace and defence industries. It is a systematic and analytical quality planning tool to identify, for new products or services at the design stage, what could go wrong during manufacture, or when in use by the customer. It is an iterative process and questions asked/examined are:

1. What is the function?
2. Potential failure modes.
3. The effect of potential failure.
4. Review of current controls.
5. Determination of risk priority (occurrence, detection and severity of failure).
6. Identification of corrective actions to eliminate failure.
7. Monitoring of corrective actions and countermeasures.

Gantt chart, *see* Bar chart

Green Belts Staff trained to be Six Sigma project leaders; they work under the guidance of Black Belts (*see* Black Belts).

Green Sigma A trademark concept by IBM to provide smart solutions for energy, water, waste and greenhouse gas management. Green Sigma applies Lean Six Sigma principles and practices to energy, water, waste and GHG (greenhouse gas) emissions throughout a company's operations.

Histogram A histogram is a descriptive and easy-to-understand chart of the frequency of occurrences. It is a vertical bar chart with the height of each bar representing the frequency of an occurrence.

Input–process–output diagram All operations or processes have inputs and outputs. The process is the conversion of inputs into outputs. Analysis of inputs should be made to determine factors that influence the process, for example input materials from suppliers meeting specification, delivery on time and so on.

Ishikawa The Ishikawa, fishbone or cause and effect diagram was developed by Kaoru Ishikawa. The premise is that generally, when a problem occurs, the effect is very obvious; and the temptation is to treat the effect. With the Ishikawa approach the causes of the effect are sought. Once the cause is known and eliminated, the effect will not be seen again. For example, working overtime is an effect, adding extra staff does not remove the cause. The question should be why is overtime necessary and what caused this problem.

ISO 9000 To gain ISO 9000 accreditation an organisation has to demonstrate to an accredited auditor that they have a well-documented standard and consistent process in place which achieves a defined level of quality or performance. ISO accreditation will give a customer confidence that the product or service provided will meet certain specified standards of performance and that the product or service will always be consistent with the documented standards.

JIT 'Just in time' was initially a manufacturing approach where materials are ordered to arrive just when required in the process, no output or buffer stocks are held, and the finished product is delivered direct to the customer. Lean Sigma incorporates the principals of JIT and now relates to the supply chain from supplier and supplier's supplier, through the process to the customer and the customer's customer.

Kaizen 'Kaizen' is a Japanese word derived from a philosophy of gradual day-by-day betterment of life and spiritual enlightenment. This approach has been adopted in industry and means gradual and unending improvement in efficiency and/or customer satisfaction. The philosophy lies in doing little things better so as to achieve a long-term objective.

Kai Sigma A combination of Kaizen and Six Sigma methodology. The approach is to mobilise resources more intensively over a shorter time horizon than traditional Six Sigma projects to yield results more quickly.

Kanban 'Kanban' is the Japanese word for card. The basic kanban system is to use cards to trigger movements of materials between operations in production so that a customer order flows through the system. Computer systems eliminate the need for cards, but the principle is the same. As a job flows through the factory, completion of one stage of production triggers the next so that there is no idle time, or queues, between operations. Any one job can be tracked to determine the stage of production. A 'kanban' is raised for each customer order. The kanban system enables production to be in batches of one.

KPIs Key performance indicators include measurements of performance such as asset utilisation, customer satisfaction, cycle time from order to delivery, inventory turnover, operations costs, productivity and financial results (return on assets and return on investment).

Lean Sigma, *see also* JIT Lean was initially a manufacturing approach where materials are ordered to arrive just when required in the process, no output or buffer stocks are held, and the finished product is delivered direct to the customer. Lean Sigma incorporates the principals of Six Sigma, and is related to the supply chain from supplier and supplier's supplier, through the process to the customer and the customer's customer.

Milestone tracker diagram Used to show the projected milestone dates and the best estimated date on the week of the progress review on a single chart.

Mind mapping A learning tool for ordering and structuring the thinking process of an individual or team working on a focused theme. According to Buzan, the mind map 'harnesses the full range of cortical skills – word, image, number, logic, rhythm, colour and spatial awareness – in a single, uniquely powerful technique'.

Minitab A statistical package developed at the Pennsylvania State University and distributed by Minitab Inc. Today, Minitab is often used in conjunction with the implementation of Six Sigma and other statistics-based process improvement methods.

Mistake-proofing Refers to making each step of production mistake-free. This is also known as 'poka yoke'. Poka yoke was developed by Shingo (*see also* SMED), and has two main steps: (1) preventing the occurrence of a defect and (2) detecting the defect. The system is applied at three points in a process:

1. In the event of an error, prevent the start of a process.
2. Prevent a non-conforming part from leaving a process.
3. Prevent a non-conforming product from being passed to the next process.

Monte Carlo technique A simulation process. It uses random numbers as an approach to model the waiting times and queue lengths and also to examine the overall uncertainty in projects.

MRP(II) Manufacturing resource planning is an integrated computer-based procedure for dealing with all the planning and scheduling activities for manufacturing, and includes procedures for stock reorder, purchasing, inventory records, cost accounting and plant maintenance.

Mudas 'Muda' is the Japanese for waste, or non-value-adding. The seven activities considered are: excess production, waiting, conveyance, motion, process, inventory and defects. For further details, see Chapter 1.

Nominal group technique A brainstorming tool for a team to come to a consensus on the relative importance of ideas by completing importance rankings in a team's final priorities.

Normal distribution (or Gaussian distribution) A graph of the frequency of occurrence of a random variable. The distribution is continuous, symmetrical, bell-shaped and the two tails extend indefinitely.

OEE Overall equipment effectiveness is the real output of a machine. It is given by the ratio of the good output and the maximum output of the machine for the time it is planned to operate.

Pareto Wilfredo Pareto was a 19th-century Italian economist who observed that 80% of the wealth was held by 20% of the population. The same phenomenon can often be found in quality problems. Juran (1988) refers to the vital few and the trivial many. The technique involves collecting data of defects, identifying which occur the most, and which result in the most cost or damage. Just because one defect occurs more often than others does not mean it is the costliest or should be corrected first.

PDCA The plan–do–check–act cycle was developed by Dr W.E. Deming. It refers to: *Planning* the change and setting standards; *Doing* – making the change happen; *Checking* that what is happening is what was intended (standards are being met); and *Acting* – taking action to correct back to the standard.

Performance charts (or UCL/LCL) Upper control and lower control limits are used to show variations from specification. Within the control limits, performance will be deemed to be acceptable. The aim should be over time to reduce the control limits. Thus, control charts are used to monitor processes and the data gathered from the charts should be used to force never-ending improvements. These types of charts might also be known as tolerance charts.

Poka yoke Refers to making each step of production mistake-free. This is known as mistake-proofing. Poka yoke was developed by Shingo (*see also* SMED), and has two main steps: (1) preventing the occurrence of a defect and (2) detecting the defect. The system is applied at three points in a process:

1. In the event of an error, prevent the start of a process.
2. Prevent a non-conforming part from leaving a process.
3. Prevent a non-conforming product from being padded to the next process.

Project A project is a unique item of work for which there is a financial budget and a defined schedule.

Project management Involves the planning, scheduling, budgeting and control of a project using an integrated team of workers and specialists.

Project charter A project charter is a working document for defining the terms of reference of each Six Sigma project. The charter can make a successful project by specifying necessary resources and boundaries that will in turn ensure success.

Process capability The statistically measured inherent reproducibility of the output (product) turned out by a process that is within the design specifications of the product.

Process mapping A tool to represent a process by a diagram containing a series of linked tasks or activities which produce an output.

PESTLE Political, Economic, Social, Technical, Legal and Environmental – an analytical tool for assessing the impact of external contexts on a project or a major operation, and also the impact of a project on its external contexts. There are several possible contexts.

QFD Quality function deployment is a systematic approach of determining customer needs and designing the product or service so that it meets the customer's needs first time and every time.

Qualitative Uses judgement and opinions to rate performance or quality. Qualitative attempts to 'measure' intangibles such as taste, appearance, friendly service, etc.

Quality circles Teams of staff who are volunteers. The team selects issues or areas to investigate for improvement. To work properly, teams have to be trained – first in how to work as a team (group dynamics) and second in problem-solving techniques.

Quality project teams A top-down approach to solving a quality problem. Management determines a problem area and selects a team to solve the problem. The advantage over a quality circle is that this is a focused approach, but the disadvantage might be that members are conscripted rather than volunteers.

Quantitative Means that which is tangible or can be measured. For example, the speedometer on a car measures and shows the speed.

Regression analysis A tool to establish the 'best fit' linear relationship between two variables. The knowledge provided by the scatter diagram is enhanced with the use of regression.

Rolled throughput yield (RTY; *see also* FPY) The ratio of the number of completely defect-free units at the end of a process (without any kind of rework during the process) and the total number of units at the start of a process. The theoretical throughput rate is often regarded as the number of units at the start of the process. RTY/FPY is used as a key performance indicator to measure overall process effectiveness.

RU/CS Resource utilisation and customer service analysis is a simple tool to establish the relative importance of the key parameters of both resource utilisation and customer service and to identify their conflicts.

Run chart A graphical tool to study observed data for trends or patterns over a specific period of time.

S&OP Sales and operations planning is derived from MRP and includes new product planning, demand planning, supply review – to provide weekly and daily manufacturing schedules – and financial information. *See also* MRP(II). S&OP is explained further in Chapter 6 (see Figure 6.10).

Scatter diagram These diagrams are used to examine the relationship between two variables. Changes are made to each and the results of changes are plotted on a graph to determine cause and effect.

Sigma The sign used for standard deviation from the arithmetic mean. If a normal distribution curve exists, one sigma represents one standard deviation either side of the mean and accounts for 68.27% of the population. This is explained more fully in Chapter 3.

SIPOC A high-level map of a process to view how a company goes about satisfying a particular customer requirement in the overall supply chain. SIPOC stands for supplier, input, process, output and customer.

Six Sigma A quality system which in effect aims for zero defects. Six sigma in statistical terms means six deviations from the arithmetic mean. This equates to 99.99966% of the total population, or 3.4 defects per million opportunities.

SMED Single minute exchange of dies was developed for the Japanese automobile industry by Shigeo Shingo in the 1980s and involves the reduction of change-over of production by intensive work study to determine in-process and out-process activities and then systematically improve the planning, tooling and operations of the change-over process. Shingo believed in looking for simple solutions rather than relying on technology.

SoQ The signature of quality is a self-assessment process supported by a checklist covering: customer focus, innovation, personnel and organisational leadership, use of technology, and environment and safety issues. It is useful in FIT SIGMA for establishing a company 'health' report.

SPC Statistical process control uses statistical sampling to determine if the outputs of a stage or stages of a process are conforming to a standard. Upper and lower limits are set, and sampling is used to determine if the process is operating within the defined limits.

The seven wastes (*see also* muda) 'Muda' is the Japanese for waste, or non-value-adding. The seven activities considered are: excess production, waiting, conveyance, motion, process, inventory and defects. For further details, see Chapter 1.

SWOT A tool for analysing an organisation's competitive position in relation to its competitors.

Tolerance charts (or UCL/LCL) Upper control and lower control limits are used to show variations from specification. Within the control limits performance will be deemed to be acceptable. The aim should be over time to reduce the control limits. Thus, control charts are used to monitor processes and the data gathered from the charts should be used to force never-ending improvements. These types of charts might also be known as performance charts.

TPM Total productive maintenance requires factory management to improve asset utilisation by the systematic study and elimination of major obstacles – known as the 'six big losses' – to efficiency. The 'six big losses' in manufacturing are breakdown, setup and adjustment, minor stoppages, reduced speed, quality defects, and start up and shut down.

TQM Total quality management is not a system, it is a philosophy embracing the total culture of an organisation. TQM goes far beyond conformance to a standard, it requires a culture where every member of the organisation believes that not a single day should go by without the organisation in some way improving its efficiency and/or improving customer satisfaction.

UCL/LCL Upper control and lower control limits are used to show variations from specification. Within the control limits performance will be deemed to be acceptable. The aim should be over time to reduce the control limits. Thus, control charts are used to monitor processes and the data gathered from the charts should be used to force never-ending improvements. These types of charts might also be known as performance charts, or tolerance charts.

Value analysis Very often a practice in purchasing; the evaluation of the expected performance of a product relative to its price.

Value chain Also known as Porter's value chain. According to Michael Porter, the competitive advantage of a company can be assessed only by seeing the company as a total system. This 'total system' comprises both primary and secondary activities.

Value stream mapping VSM is a visual illustration of all activities required to bring a product through the main flow, from raw material to the stage of reaching the customer.

Visual Six Sigma Because of its visual emphasis on Six Sigma tools, Visual Six Sigma facilitates a rapid exploration of the data to quickly drive focus onto the root causes. The modelling and simulation capability then builds on that to allow further testing of 'what-if' scenarios and investigation of different improvement options.

World class The term used to describe any organisation that is making rapid and continuous improvement in performance and is considered to be using 'best practice' to achieve world-class standards.

Zero defects Philip Crosby made this term popular in the late 1970s. The approach is 'right thing, right time, right place, and every time'. The assumption is that it is cheaper to do things right the first time.

Index